O9-BUD-720

THE SPY WHO KNEW TOO MUCH

Also by Howard Blum

NONFICTION

Night of the Assassins

In the Enemy's House

The Last Goodnight

Dark Invasion

The Floor of Heaven

American Lightning

The Eve of Destruction

The Brigade

The Gold of Exodus

Gangland

Out There

I Pledge Allegiance

Wanted!

FICTION

Wishful Thinking

Pete Bagley, before he entered the secret world.

THE SPY WHO KNEW TOO MUCH

An Ex-CIA Officer's Quest Through a Legacy of Betrayal

HOWARD BLUM

HARPER
An Imprint of HarperCollins*Publishers*

HarperCollins books may be purchased for educational, business, or sales promotional use. For information, please email the Special Markets Department at SPsales@harpercollins.com.

FIRST EDITION

Library of Congress Cataloging-in-Publication Data has been applied for.

ISBN 978-0-06-305421-9

22 23 24 25 26 LSC 10 9 8 7 6 5 4 3 2 1

For Phil Werber,

My good buddy since way back when

And Leonard Novins,

In memory

There's no such thing as a former spy.

—A KGB wisdom

On the second day, a sail drew near, nearer, and picked me up at last. It was the devious cruising Rachel, *that in her retracing search after her missing children, only found another orphan.*

—Herman Melville, *Moby-Dick*

Contents

A Note to the Reader

IN THIS BOOK, MY INTENTION is to reveal one of the last great secrets of the Cold War. It is also the true story of one spy's quest through a legacy of betrayals to solve this mystery. And to accomplish both these goals, I relied on not only previously classified government documents, memoirs, interviews with both present and past officers in the intelligence services but also, most helpfully, conversations with individuals who were intimately connected to the characters who animate this story. A complete chapter-by-chapter sourcing appears at the end of this book.

But let me also share another advisory: When the hero of this story, Pete Bagley, knew he was dying, he wrote to a friend, "In the future an alert journalist or historian, inspired by some new revelation, may remember one or another of these old ghosts and dig deeper to lay them to rest."

And that's what I set out to do in the pages that follow.

—HB

Cast of Characters

(In Order of Appearance)

The Americans

TENNENT "PETE" BAGLEY: Counterintelligence officer and deputy head of the CIA's Soviet Bloc division.

MARTI PETERSON: The first female case officer assigned to Moscow Station.

JOHN PAISLEY: CIA analyst with a wide-ranging portfolio, which included defector interrogations as well as Soviet military strategy and nuclear weapons capabilities.

RAY ROCCA: Head of Research and Analysis, the CIA's Counterintelligence Staff.

JAMES ANGLETON: Chief of the CIA's Counterintelligence Staff.

CLARE EDWARD PETTY: Member of CIA's Special Investigative Group (SIG).

CHRISTINA BAGLEY ROCCA: Pete Bagley's daughter, a CIA officer who married Gordon Rocca, a DIA analyst and the son of Ray Rocca.

WILLIAM COLBY: CIA fieldman who became director of Central Intelligence.

GEORGE KISEVALTER: Russian-born CIA officer who served as handler for several double agents.

JACK MAURY: CIA Soviet Division chief.

WILLIAM HOOD: Cold War Vienna Station chief.

DAVID MURPHY: Berlin fieldman and CIA Soviet Division chief.

RICHARD HELMS: Wartime OSS officer who rose through the ranks to become CIA director.

JOHN ABIDIAN: Security officer at the American embassy in Moscow who performed operational tasks for the CIA.

BRUCE SOLIE: CIA security officer who defended Nosenko's bona fides and later played a key role in the ill-fated running of double agent Nicholas Shadrin.

LEONARD MCCOY: CIA reports officer who defended Nosenko, asserting he was not a dispatched Russian agent.

JOHN HART: CIA officer who cleared Nosenko and later gave testimony to the House Select Committee on Assassinations that was pointedly critical of Pete Bagley.

KATHERINE HART: Chief of staff for CIA field stations and wife of John Hart.

Maryann Paisley: Wife of John Paisley and, for a time, a CIA clerk working directly for Katherine Hart.

David Sullivan: CIA analyst who leaked information to an aide of a US senator and later reported his suspicions about John Paisley to the Office of Security.

The Russians

Alexander Ogorodnik: Double agent code-named Trigon who, when caught, committed suicide by ingesting a cyanide pill concealed in a fountain pen.

Pyotr Popov: Lieutenant colonel in military intelligence (GRU) who provided military secrets to the CIA and was executed for treason.

Oleg Penkovsky: Colonel of GRU who passed secret intelligence to both the CIA and MI6 and was executed for treason.

Leonid Brezhnev: Soviet general secretary whose private conversations were covertly recorded in the course of the CIA's Gamma Guppy operation.

Boris Nalivaiko: KGB officer based in Vienna who lured the CIA into an embarrassing trap.

General Oleg Gribanov: Chief of KGB counterintelligence (Second Chief Directorate) who established a special unit to focus on "operational deception."

Lieutenant General Sergey Kondrashev: High-ranking KGB officer with a wide-ranging career in foreign intelligence and counterintelligence operations.

The Poles

Michal Goleniewski: Polish intelligence officer code-named Sniper.

The Czechs

Karl and Hana Koecher: Husband-and-wife team of Czech intelligence officers who worked closely with the KGB and succeeded in infiltrating the CIA.

The Defectors

Peter Deriabin: KGB officer who became a consultant to the CIA.

Yuri Ivanovich Nosenko: KGB officer who defected after the Kennedy assassination.

Anatoly Golitsyn: KGB officer who later worked closely with the CIA's Counterintelligence Staff.

Igor Kochnov: KGB agent who pretended to be a defector in order to set a trap for a Russian-born American citizen working for the DIA.

Nicholas Shadrin: Soviet naval captain who became a US double agent, an operation that resulted in his kidnapping and death at the hands of the KGB.

THE SPY WHO KNEW TOO MUCH

Prologue:
The Weight of Guilt

GUILT IS A HEAVY BURDEN. It weighs down on the heart, an unremitting punishment. Yet she did not try to escape the pain, or find excuses to wiggle out of the blame.

Instead, she acknowledged her complicity. There were things she might have done that could have made a difference. That's the definition of guilt, she discovered: knowing all you should have done.

It didn't matter that she had not been in the car that night. It didn't matter that the accident occurred at the exit that led to the main entrance of the Central Intelligence Agency. Or that this was where her husband worked. And where she had worked, too.

All that mattered was that a young, handsome boy, her son's best friend, had been killed.

In the terrible aftermath, she'd blamed her husband, too. It hadn't been his fault; he had played no role in the night's heartbreaking events. And yet! She knew, as any wife would know, that he'd created the reckless world that had inevitably bred this tragedy.

Full of rage, raw with shame, after nearly two decades of marriage, she had demanded a divorce. And with her anger, she had driven him into the arms of her best friend.

She now saw that it was all her own doing. As things had become undone, she'd capriciously kept yanking the dangling threads. And in the end, her life had unraveled.

Her punishment: a ceaseless, unabated guilt.

But all her guilt was nothing, no, less than nothing, when measured against the pain caused by the new, sinister knowledge that had taken hold of her life. It had the power to change everything that had come before, to turn long-accepted truths into lies. It was a very dangerous secret.

But she knew it could not be shared. She did not dare. It must be entombed forever in the shrewd armory of her heart.

Part I

“Once More unto the Breach”

1977–1983

Chapter 1

Brussels, October 1978

TWO DEATHS—EACH PURPORTEDLY A SUICIDE, each with its roots deep in the secret world, each with its own perplexing mysteries—wrenched Pete Bagley, retired and somewhat besmirched spy, from the complacency of his pleasant exile and set him on the twisting path back to the shadowy battlefields of his previous life. It would be, he fully recognized, his final mission, his last chance to set straight the betrayals, both personal and professional, that had scarred not just the agency, but also his own family of spies. And like every old man who at last musters the courage to confront unfinished business, he could only hope that it was not too late.

THE FIRST DEATH HAD OCCURRED without Pete's—he'd been christened Tennent, but his mother early on had started calling him Pete and the name stuck—immediate knowledge; at the time he'd been living a retiree's life of contemplative leisure with his wife and his books in the pretty city of Brussels. In fact, the suicide—if that was what had really happened—had been kept a closely guarded secret, and it wasn't until a month or so had passed—the shared time line was deliberately murky—that the Soviets allowed the grim news to leak. Of course by then, in the aftermath of the menacing arrest and the ensuing diplomatic blowup, there was no longer any operational reason for secrecy. Still, the battle-scarred cold warriors in the SB, as the CIA's Soviet Bloc Division, where Pete had once served as deputy chief, was known, couldn't help but wonder if the normally reticent KGB hoods had only grown talkative because they couldn't resist giving the knife they had planted deep into the heart of Moscow Station another vindictive twist.

The least disputed parts of this drama began to play out on the evening of July 15, 1977, a deceptively calm and quiet summer's night in Moscow. A cooling breeze floated off the Moscow River, party apparatchiks hurried across Red Square on their way home from work, and lovers abandoned the pedestrian bustle of Prospekt Mira to disappear hand in hand into the secluded nooks of the Apothecary Garden. But it was a time of high alert in Moscow Station. In the boxlike seventh-floor spook's nest hidden away in the US embassy on Novinskiy Boulevard, the spies were hoping that tonight's operation would calm their worst fears.

When an agent goes silent, there can be many benign reasons. Operatives, too, have their quotidian, overt lives to live. They can catch the flu, grow frazzled trying to placate the demands of a relentless boss, get roped into entertaining visiting in-laws, or even mark the wrong date on their calendars. But hard-nosed professionals wearily concede that the search for excuses is largely wishful thinking. When a usually productive Joe can't be contacted, when he misses a rendezvous or doesn't service his dead drops, the truth is staring you in the face with a sickening inevitability: He's been compromised, no doubt languishing in a cell in the Lubyanka, if he hasn't already been summarily dispatched by a firing squad.

Tonight would bring clarity. It would resolve once and for all the disquieting questions surrounding the all-star agent the station was running deep inside the enemy's citadel—the spy code-named Trigon.

Four years earlier, in a steamy Turkish bath in Bogotá, Colombia, a CIA officer with only a towel wrapped tight around his waist for a semblance of operational propriety had sidled up to Alexander Ogorodnik, a silky midlevel diplomat at the Soviet embassy, and had launched into his recruitment pitch. There's no transcript of what was said, but presumably the CIA recruiter would have first methodically recounted all the reckless behavior that in the course of just a brief posting had characterized the married foreign service economist's very undiplomatic life in Latin America; e.g., his frenetic juggling of romances with several of his colleagues' wives, his illegal wheeling and dealing of automobiles purchased at diplomatic discounts, his teetering pile of debts, and, not least, the recent announcement by his young Colombian mistress that she was pregnant. The implication would've been clear: If the Americans without really trying had discovered all this, how long would it take for the diplomat's Soviet comrades to catch on to his shenanigans? And it would've been unnecessary to point out that the dour Russian foreign service bureaucrats were as unforgiving as they were judgmental. Then when the CIA man saw that it wasn't just the steam that was causing the diplomat to sweat, he'd munificently offer up a way out of all this mess: a chance to earn the sort of money that would make old problems go away, as well as finance new ones.

The moments that follow any approach are always pregnant, a tense time when things might veer off in any direction. There's no telling if the prey will scream with indignation, or if he'll slink off shamefaced to put a bullet in his head. However, Ogorodnik, according to the bemused CIA accounts, didn't hesitate. He promptly announced that he'd never been a fan of the Soviet system. In fact, he insisted, he'd always been a capitalist at heart. And as if to prove it, he quickly proposed a very lucrative arrangement for his services.

A crash course in tradecraft was conducted over several weeks in a room in a Hilton hotel in downtown Bogotá. Shooting documents with a tiny T50 camera concealed in a fountain pen as well as mastering the protocols for dead drops can be a tricky business. Yet to his new handlers' delight, the diplomat was a natural. With surprising speed, the agent code-named Trigon was up and running.

Only it wasn't long before disappointment set in. The camera work was first-rate, the deliveries flawless, but top dollar was being shelled out for bargain-basement product. The spymasters in Langley had a bad case of buyers' remorse.

Then in 1974, Ogorodnik was transferred back to Moscow and given a desk in the Ministry of Affairs that gave him access to a steady stream of top secret memos and planning documents. And just like that, the CIA's high-priced investment turned prescient. Trigon was soon making regularly scheduled drops of rolls of T50 films that, when developed, brought a treasure trove of secrets into focus. Moscow Station now had eyes in the enemy's house. Langley was head over heels.

But after nearly two high-flying years, the exuberant mood had grown subdued, even a bit glum. Warning signs had begun to appear. In January 1977, a CIA officer skied through a fresh blizzard of snow to the designated drop site in a forest on the outskirts of Moscow. Nothing could be found. Perhaps the snow had deterred Trigon, the optimists wanted to believe. So after an uneasy month of waiting, Moscow Station tried again: A hollowed log filled with previously requested communication gear was left at the usual site. It was never retrieved. But then in April, Moscow Station broke out in cheers when Ogorodnik left a cache of film canisters as scheduled. Only once the station's tech officers sorted through the material, there were renewed doubts. Every spy has his own handwriting, the way he goes about his clandestine tasks, and to the discerning eyes of the analysts at the embassy, this cache didn't seem to be Trigon's handiwork. It was too sloppy, assembled without his usual meticulous tradecraft.

Yet refusing to accept the unacceptable, Moscow Station decided they'd give Trigon one more try. A coded message was sent by shortwave radio requesting that if he was ready to resume work, he should send the prearranged signal.

And to the rekindled excitement of the true believers in Moscow Station, he did. A red dot appeared on a "Children Crossing" traffic sign adjacent to a Moscow school.

There were naysayers, though, who had misgivings. They dismissed this signal as a lure. To their skeptical eyes, the red dot was clearly stenciled—and no genuine fieldman would stick his head out of the shadows long enough to execute that sort of painstaking procedure. Further, the stenciled dot was colored in a red as bold and bright as the Soviet flag—and that, too, didn't seem a secret agent's furtive doing. It was all too deliberate; in the field, subtlety was the guiding rule.

In the end, both sides agreed there was only one way to find out for sure.

UNBUTTONING HER BLOUSE, THE SPY attached the tiny radio receiver to her bra with a Velcro tab. Her long, streaked blond hair hid the earpiece. If the KGB watchers were tailing her, she'd now be able to eavesdrop on their transmissions. But that, she realized, offered only small reassurance. If the opposition was on to what was going down, that meant it was already too late. For Trigon, and for herself.

It was just after six on that July evening in Moscow when Marti Peterson, a willowy thirty-two-year-old and the first female case officer ever assigned to Moscow Station, left her apartment and headed off for the drop. She clutched a bag containing what looked like a lump of black asphalt. A closer examination of the shard, however, would reveal a secret compartment; inside were messages and a new, improved miniature camera that the tech wizards at Langley had fabricated just for Ogorodnik.

As Peterson got behind the wheel of her squat Zhiguli, she wanted to believe that all would go well tonight. The sexist dinosaurs at the KGB, she'd been assured by her gung-ho station chief, would never suspect that the agency was running a female officer in Moscow, let alone a young and attractive one. And after two cautious years in the city, she, too, remained convinced that her cover had not been blown; to the First Department watchers who kept a vigilant eye on embassy personnel, she remained just another clerical worker, a woman of no intelligence interest. But Peterson also knew that the KGB routinely blanketed the city with a small army of its operatives. They had roving cars, pavement artists, as well as hidden cameras all over Moscow. And on a summer's evening when the sun stayed high in the sky till absurdly late, when even at ten p.m. the shadows would only be gloaming, the watchers would have nature on their side, too.

With the careful, well-practiced discipline of an agency professional, Peterson went through the maneuvers to shake any tails. She drove around the city, turning left and right at whim, her eyes darting to the rearview mirror. When she was convinced that no one was following, she parked and headed to the subway. She rode to the first stop, then switched lines, traveling now in the opposite direction. Studying the faces reflected in the train window, she looked for a sign that she'd been targeted. But even as she decided there was no cause for alarm, Peterson knew that if the A team was out tonight, they'd know better than to stare.

Peterson got off at the sports stadium at Luzhniki, and, luck on her side, a soccer match had just ended. She floated along in the sea of departing spectators and let the crowd carry her forward, until once again she was assured that no was paying any attention to her. Then she slipped out of the swarm and hightailed it to the drop point, arriving with a well-trained sense of timing at precisely the prearranged tick of 10:15.

An ancient stone tower, crenellated like a turret from a medieval fortress, rose up from a railroad bridge that spanned a lonely stretch of the Moscow River. The stairs inside the tower were slippery, the old stone worn with age, and she trod with care. Peterson silently counted forty steps and then looked up to find the casement window where she'd been told it would be. Starlight seeped through, casting opaque shadows on the thick, dark walls. She placed the asphalt shard exactly one arm's length from the windowsill. No one would notice it, unless, of course, they knew where to look.

Her instinct was to rush down the stairs, to get away as quickly as possible, but with a leisureliness that was all discipline she descended slowly. One step after another, she made her way. The space was as tight as a cocoon; she felt trapped, there'd be nowhere to flee if she had to run. In her desperation, she listened for stray night sounds.

As she exited the tower, three men rushed toward her. She saw them coming; their shirts were very white against the backdrop of the river, the old gray stones, and the incipient night. At that horrible moment, a number of thoughts raced as if in a single instant through her mind. The hoods had no need to follow her because they knew where she was going even *before* she'd left her apartment. That meant Trigon had been burned. Yet she couldn't be sure, so she needed to warn him to stay away. At the top of her lungs she heard herself yelling, "Provocation! Provocation!" And even as she screamed, there was a wild moment when she considered jumping into the river, executing a swan dive like a spy in a Hollywood movie. But at the same time she decided it'd be a doomed escape; if she survived the impact, they'd send out boats to scoop her up.

All at once they were on her. One of the men ripped her blouse, searching for the wire, and he let his hands linger, enjoying the hunt. Three against one, and she knew there was no way to spring free, but her rage was fueled by the intrusiveness of the assault. She fought back wildly; hands balled into fists rained blows. They grabbed her arms, fixing them tightly to her sides, so she let loose with a kick. It landed with power, straight into a groin, and her victim fell to the ground with an agonized scream.

But now a van had pulled up and more of them piled out. She was surrounded, and someone started taking photographs, the flashbulb going off like tracers in the night. "Let me go!" she yelled. She could see they'd retrieved the hunk of asphalt, and one of the Russians had raised it above his head, rejoicing like an athlete who'd just scored a goal. She insisted that it was all a mistake. Call the

American embassy, she exhorted. But she knew like any professional would that there was no way out.

The van took her to the Lubyanka, and they led her handcuffed across a long concrete courtyard. Peterson feared that she'd be locked in a cell for a few days, and the thugs would work on her until she signed a confession. Instead, a team of interrogators put on an elaborately orchestrated show. She was seated at a table in a windowless room, and with the cameras rolling to preserve every incriminating moment, the asphalt chunk was brought in. The headman, as self-satisfied as a magician pulling rabbits out of a hat, began emptying the contents from the hidden cavity: a coded message imprinted on 35 mm film; a black fountain pen that concealed a camera; and, spoils for the traitor, tight rolls of rubles and emerald jewelry. Peterson's face remained rigid, betraying nothing.

Later that night she was released, and a stolid American consular officer drove her straight to the embassy. The universal frame of mind at Moscow Station was one of despair, but the debriefing had to be done while Peterson's memory was fresh. At three thirty that morning, an enciphered summary of the debrief was sent as a Flash signal to Langley. Peterson flew out of Moscow later that day, officially branded a persona non grata. If she'd been asked, she'd volunteer that the feeling was mutual.

And now the waiting began. Yet the only question that remained was how bad the news would be.

It couldn't have been worse. There were two versions of Ogorodnik's fate, both of them leaked over the next couple of months through semiofficial Soviet channels, and while the details of what had happened on the night of June 21, 1977—not quite a month before Peterson was sandbagged—differ, the woeful punch lines were identical.

One grisly account had a tough-guy squad led by a general from the KGB's Seventh Directorate pounding down the door of Trigon's apartment on Krasnopresnenskaya Embankment. The invaders swiftly uncovered the cached CIA communications equipment, and, obeying the Kremlin's predetermined decision that a trial would be an embarrassing formality, they executed Ogorodnik on the spot.

The other version gave the captured spy a more heroic finale. Here, too, the spy was rumbled in his apartment. Only in this telling, he's forcibly stripped to his underwear and a no-holds-barred interrogation begins. They want his contacts, his drop sites, whatever else he knows about the enemy. Ogorodnik, despite growing increasingly worse for the wear and tear, won't give an inch. But finally he breaks, just as anyone eventually would under similar duress. Just stop, he begs. I'll tell you everything. Give me my pen, he offers with a weary resignation, and I'll write it all out. So they handed him the black Parker pen from his desk, and suddenly he bit down hard on the barrel and ingested the concealed cyanide pellet. He was dead before his astonished interrogators could wrest the pen from his lifeless hand.

Of course, there was still the mystery of how Trigon was blown. The stories coming out of Moscow Center helpfully cleared that up, too. The credit, the spymasters boasted, belonged to the counterintelligence owls in the Second Chief Directorate. They'd noticed Ogorodnik's car parked time after time in the same spot near Victory Park and then deduced that this had all the markings of a prearranged signal; the location was on the route many employees at the American embassy would normally pass on their way to work. Once they'd caught the scent, they maintained surveillance until Ogorodnik was observed emptying a dead drop. And if anyone didn't believe that bit of detective work, dismissed it as too much of a coincidence, well, there was a highly publicized confirmation:

The leader of the Trigon arrest team, a Second Chief Directorate officer, had, with much ceremony, been awarded the Order of the Red Banner for his work on the case.

BUT AFTER THE BROAD DETAILS of this debacle and the subsequent postmortem rationalizations made their meandering way to Pete in Brussels, he sat in his book-lined study and tried to sort it all out. His outsider's instinct was not to pin the blame on either slapdash tradecraft or the malicious efficacy of bad luck. Instead, his thoughts broke out into a familiar sort of mayhem, racing back into the past with anger and concern. He knew now, just as he had known back then. He was certain. And the old fear rose up in him again.

Chapter 2

But it was the second death that hardened Pete's certainty, and in that confirming way further challenged him to make a decision. It occurred a little more than a year later, and the events were played out not behind enemy lines but rather in the homeland from which he'd fled. And the fact that this suicide, as it was also being called, occurred at sea, well, that just gave the mystery a more piquant pull; after all, he had been a sailor's son way before he had been a spy.

Pete came from a long line of old salts, each a navy man of formidable and distinguished reputation. A small flotilla of warships, from frigates to cruisers, had been christened with the names of his father and uncles. And while his two brothers, in keeping with the family tradition, had shipped off to the US Naval Academy and gone on to become the first siblings to sport four stars on their shoulders, Pete's dodgy eyesight had prevented him from becoming a midshipman. So, breaking with familial convention, he headed to Princeton at a precocious seventeen. But when the war came, he quickly enlisted in the US Marines and, as if fated, wound up at sea as a lieutenant in a marine detachment on an aircraft carrier. And with the arrival of the new Cold War, and after putting in a couple of brief stints of active duty as a Marine Reserve officer, it was the judicious shaking of the well-connected branches of his naval family tree that conspired to ease Pete's path into the secret world.

The year was 1949, and Pete, taking advantage of the GI Bill, had been spending long days toiling at a desk in the Library of

Congress researching a high-minded dissertation on the intricacies of nineteenth-century diplomacy. His much more scintillating nights, however, were spent bivouacked at his uncle Bill's pleasant house on Florida Avenue. One night at dinner, Pete, having come to the realization that academic life promised only a humdrum future and, to boot, no opportunity to be of genuine service to the nation, mentioned that once his doctorate was completed, he was mulling applying to the newly created Central Intelligence Agency.

"Good idea," agreed Uncle Bill, who also happened to be Fleet Admiral William D. Leahy, the very man who had been one of the founding fathers present at the creation of the Central Intelligence Group, the CIA's immediate postwar predecessor.

"I'll mention it to Hilly," Uncle Bill added helpfully. "Hilly," to those on less intimate terms, was Rear Admiral Roscoe Hillenkoetter, the director of the CIA. They went back a ways; when Bill had been ambassador to unoccupied France in 1940–41, Hilly had been the naval attaché. These postwar days, Uncle Bill was also known around DC as one of the men who had President Truman's ear, and that didn't hurt either.

With a swift "Aye, aye, sir," Pete's CIA application was stamped "approved."

Now a lifetime of secrets later, after all his frontline Cold War adventures, Pete found himself on the outside looking uncomfortably back at a clandestine world from which he had walked away. And it wasn't just the maritime connection that gave this second baffling death—murder? suicide?—such a strong pull on his emotions. The corpse was someone he had known a bit, a man with whom he'd done some things in the shadows.

AT FULL SAIL, ITS BIG white sheets spread by the crisp autumn wind, the *Brillig* was a sight to behold, the sloop majestic as it glided through a Chesapeake Bay glistening in speckled early-morning sunlight. It was just after nine, a fresh Monday, September 25, 1978. Robert McKay, an ancient mariner who had made his living for seemingly eons crabbing in the waters off his hometown port of Ridge, Maryland, watched the approaching craft from the weather-beaten deck of his *Miss Lindy* with a heartfelt pang. "It looked so pretty, like it was in a race," he mused with a sailor's covetous admiration.

Then it was as if he woke with a start. All at once he realized: The sailboat was speeding straight toward him, hell-bent, and on a collision course. It must've been doing seven knots.

Moving with furious energy, while all the time screaming invectives at the deranged skipper of the rapidly approaching sloop, he began wildly spinning the wheel of the *Miss Lindy* to port as fast as he could. Only the *Brillig* kept coming, charging at him. He braced for the moment of impact.

In the end, the *Brillig* came *this* close, but somehow managed to fly by without contact. Yet as it passed, in a startling moment that put a sudden brake on his towering rage, McKay saw that there was no one at the tiller. In fact, there was no one at all on the *Brillig*'s deck.

Was the captain down below? Asleep? Hungover? Sick? Dozens of possibilities filled McKay's head.

For forty-five apprehensive minutes, McKay, full of an old-time seaman's sense of foreboding, followed the *Brillig*. His calls to the sloop went unanswered, futile volleys of concern swallowed up by the far reaches of the bay.

But when the wind changed direction, the sailboat tacked in response. Its sails still unfurled, it headed swiftly toward the shore.

The thirty-one-foot *Brillig* had a deep keel; it's what kept her steady when powerful gusts raged. But on this morning it saved her from being wrecked. As the sloop raced closer to shore, as it seemed a sure thing that she'd smash into the jagged rocks rising out of the water just beyond Hays Beach, layers of Chesapeake Bay mud trapped her deep keel like a vise. The ooze held the sailboat tight in its grip, quite effectively slowing the *Brillig* until at last, as if spent by its exertions, it came to a weary halt.

The Coast Guard station perched on the promontory overlooking Moll's Cove, that'd be the closest, McKay figured. He reached for his radio.

When you signed on for the job as a Maryland park ranger, the bosses had a habit of reminding that you need to be, just as the training manual declared, "a generalist." Ranger Gerald Sword, posted at Point Lookout State Park, liked that about the work, that it wasn't just endless days of staring out across the blue expanse of the bay toward the horizon. In his time, he'd rushed sun-worshipping summer tourists with third-degree burns to the hospital, tracked down lost toddlers, even had broken up a fistfight when the annual Civil War reenactment near the site of the old Confederate prison camp got a bit too realistic. And so when not too long after his shift had started, maybe 10:25 a.m., he received a call from the Coast Guard to check out a report that a sailboat had gone aground about two miles north of Scotland Beach, he hurried off. Only, Sword would later confess, all his generalist years had not prepared him one bit for the challenges and deep perplexities of what he stumbled into on that bright, ripe autumn morning.

"Ahoy, *Brillig.* Anyone on board?" the ranger shouted. He was standing on the beach; the grounded sloop, sails unfurled, was, he estimated, six feet from the water's edge. There was no response. Another concern: There were no footprints in the sand; no one had gotten off the boat.

He climbed onto the deck, and it was slippery beneath his feet. Looking down, he noticed the unspent cartridges—bullets from a handgun? maybe a nine millimeter, he guessed—scattered about the deck as though they'd been tossed like jacks in a children's game. Making his way aft, he saw that the self-steering gear on the wheel was engaged. That would send the boat in circles. What'd be the point? he wondered. He looked for signs of life and saw none.

The cabin door opened to his pull, and he descended. It was pitch dark, the curtains drawn, and so he spread them apart. In the shower of daylight, he saw the disarray. Papers were thrown about. The galley table had been broken. Remnants of a meal—pickle loaf, he determined with a quick sniff—littered the floor. Was this just the housekeeping of a slovenly captain? Or had there been a fight? Ranger Sword filed those questions away for later. The fact that there were no bloodstains, neither on the walls nor on the floor of the cabin, allowed him to tell himself that this would all have a happy resolution.

He turned his attention to the papers, and, willing his pounding heart to a steadier beat, he began to sort through them methodically. He hoped they'd provide some clue to the identity of the boat's owner. There were lots of pages, and as he continued, he perused with increasing concern. As best he could make out, they were some sort of government reports; in the margins, notes had been scribbled in pencil as if by a student cramming for an exam. Stray words caught his eye: missiles, Soviet Bloc, satellite. At the top of many of the pages was a stamped advisory that riveted his attention like a warning beacon: CLASSIFIED, TOP SECRET.

Beneath the pile of papers there was a small metal phone directory; slide the lever to the desired letter of the alphabet, and it pops open. Still foraging for clues to help identify the boat's owner, the ranger began trying letters at random. It seemed, he decided quickly, that most of the telephone numbers had a 351 prefix. That wasn't a local area code, he knew. Maybe the owner lived out of state. Perhaps that'd explain things: A vacationing sailor not accustomed to the vagaries of the Chesapeake's currents had stumbled overboard.

Continuing his search of the cabin, Sword's eyes riveted on an open leather briefcase. He extracted a letter. It was addressed to John A. Paisley, Washington Post Agent #1401, P.O. Box 9355,

1500 Massachusetts Avenue #847, Washington, D.C. The writer was angry, complaining that deliveries of the *Post* had been erratic, but payment was enclosed for what had been received.

Well, the ranger told himself with a small burst of investigative pride, he now had a name and an address: John A. Paisley. But no sooner had he settled on this deduction than he realized it was quite improbable: A guy who delivers newspapers owns a fancy sailboat like this? That didn't seem likely. Still, he had a name and he'd better share it. And there was a boat that needed to be towed in.

He was about to leave the cabin when, for the first time, he realized that the ship-to-shore radio had been operating, emitting small, weak squawks the entire time. He chided himself for not hearing this before, and, searching for an excuse, he decided the noise had been drowned out by the intensity of his concentration.

Twisting the dial to the off position, he noticed the carefully arranged row of electronic consoles spread across a lower shelf. The ranger was no expert, but he felt these devices were something out of the ordinary, not your everyday ship-to-shore transmitting equipment. Why have this kind of sophisticated gear on a sailboat? Why, for that matter, was a collection of expensive electronics on a boat that apparently belonged to someone who delivered copies of the *Washington Post* for a living? He added one more question to the list of puzzlements that had been filling his mind since he'd boarded the *Brillig*.

Then he hightailed it to his jeep and drove across the dunes at breakneck speed to the nearest house to phone the Coast Guard.

Chief Petty Officer James Maxton was a large, red-faced bull of a man with a furious temper. And this morning, as soon as the *Brillig* was towed to the tiny Coast Guard station he commanded at St. Inigoes, an appraising look at the fancy craft ignited his anger. Over his many years on the bay, he'd had his fill of hapless weekend sailors, and to his mind, despite Ranger Sword's feverish conundrums, what had happened on the *Brillig* was not anything out of the ordinary. It was, he decided with a conviction reinforced by his mounting rage, just another case of some self-indulgent amateur with more money than brains or seafaring skills. Now he'd have to spend his day, as he'd done so many times before, tediously cleaning up this mess. Probably turn out that John Paisley was at this very moment drunk as a skunk in some seaside bar, not even realizing that his sleek boat had sailed off without him. He didn't need to know John Paisley to know that he didn't like him. Just thinking about him, and all the other feckless mariners like him he'd encountered over the years, sent his blood boiling.

But the chief knew the drill, and, to his credit, he followed it to the letter. He called the Maryland Natural Resources Police to report that a sailor had vanished from his boat. They dutifully passed the news on to an officer at the Maryland State Police who, as if he'd been roused from a deep sleep, dully replied that he'd list the incident as a possible drowning.

And there things might have remained for at least a few more indolent days if Chief Maxton hadn't worked himself up into such a state. He wanted to give the careless Paisley a piece of his mind, but first he had to find him. So, following up on Ranger Sword's initial discoveries, he began to make inquiries.

Apparently John Paisley worked as a newsagent at the *Post*, and he started there. A while passed before Joseph Haraburda, the circulation manager of the paper, got back to him. There was no John Paisley employed at the *Post*, he insisted. Not as a newsagent or in any capacity. In fact, he went on, flaring with high dudgeon, the identification number that the chief had shared belonged to someone else, a longtime employee named Archie Alston. And Alston, well, he'd never heard of Paisley and had no idea why this stranger would be claiming his ID number.

With that, Chief Maxton's mood changed. His anger gave way to a fertile curiosity. *A metal phonebook, one of those old-fashioned things*, he recalled the ranger saying, and he went off to hunt it down.

All those telephone numbers with the 351 prefix were possible clues, he reasoned as he searched through the directory. He considered calling one, but instead, still tentative, he dialed the operator and asked what state had that area code. None, he was officiously informed. There was no such working prefix.

That threw him completely. Crazy notions started to run through his head. Who was this John Paisley? There was definitely something very odd about the entire incident. His suspicions were on high alert, but any reasonable explanation was beyond him.

Once again, he wanted to take the initiative, ring a number in the metal phonebook and see who answered. A lifetime in the Coast Guard, however, had drummed into him the necessity of protocol. There was a way things had to be done. He called headquarters in Portsmouth, Virginia, and, reining in his more outrageous theories, he shared what he knew with a Lieutenant Murray. His report was a model of objectivity, a recitation of the facts and nothing more.

Probably nothing to this, judged the lieutenant evenly. Just another boating accident. But stand by, the lieutenant instructed. I'll run this up the chain of command and get back to you.

An hour, maybe even two, had passed before headquarters called back; time had seemed to lose its meaning for the chief that disconcerting day. It was Lieutenant Murray again, only now he'd abandoned the idea that this was simply another run-of-the-mill maritime mishap.

The 351 prefix is a government exchange, he explained cryptically. "Highly classified," he added. But then, as if the secret was too juicy not to share, he blurted it out: "It belongs to the CIA."

He ordered Chief Maxton to post an armed guard on the *Brillig.*

The CIA was diligent. Two officious men from the agency's Office of Security had, over several painstaking days, combed the *Brillig* from stem to stern. By the time they were done, they had boxed the reams of documents, the battered leather briefcase, the classified phone directory, and the sophisticated electronic equipment for transport to Langley. Whenever Coast Guard or state police officers dared to challenge their jurisdiction, the two CIA men offered a terse explanation that national security issues took precedence. Pressed for further details, the agency's team refused to budge. "Need to know," they barked, effectively slamming the door on further conversation.

John Paisley's wife, Maryann, had also been informed; a resourceful Coast Guard officer had tracked down the address of her ranch home in McLean, Virginia. She had been separated from her husband since August, but, accompanied by her teenage son Edward, she immediately drove to John's bachelor apartment in downtown Washington. She hoped to find something that would give her a clue as to what had happened on the *Brillig.*

When she opened the door to the apartment, she discovered that it had been turned inside out. Someone—or someones; a team perhaps, she guessed by the extent of the devastation—had wreaked havoc. It seemed to her that they'd been looking for something, something specific. It had probably been the work of the agency's Office of Security, but when she inquired, she couldn't get a straight answer. Still, she kept her temper in check.

It wasn't simply that, after more than twenty years of marriage, she had tacitly acquired a practiced restraint. Maryann Paisley had worked at the agency, too. She held a classified clearance, was privy to many secrets, and knew from her own experience how business was conducted in the clandestine corridors of power. An obliging coconspirator, she didn't complain about the breaking and entering. And she didn't run to the press to share the news that her husband, the spy, was missing. She knew to keep her head down.

But then, a week later, a body was found.

THE DAY WAS GRAY AND gloomy, a warm drizzle falling. On Sunday, October 1, 1978, three fishermen aboard the *Miss Channel*

Queen looked beyond their lines and, like a cautionary chorus in a Greek drama, suddenly shouted out in more or less unison: body ahoy!

Within an hour a Coast Guard cutter sped to the waters east of the mouth of the Patuxent River and lowered a wire basket to trap "the floater." The corpse hauled on board was gruesome. Decomposition and the greedy nibbling of Chesapeake Bay crabs had rendered the body beyond recognition—a largely skinless, pasty specter of a human.

Yet even in the corpse's macabre state, two abnormalities immediately caught the sailors' attention. Above the left ear, a hole had been singed into the skull; brain matter continued to leak onto the deck. And, no less a cause for speculation, two sets of diver's belts—later determined to be nineteen pounds each—were wrapped around the corpse like the tight bands encasing a sarcophagus.

Alerted by radio, the county coroner was waiting when the Coast Guard cutter arrived at the Naval Center on nearby Solomons Island. "Foul play," Dr. George Weems summarily decided after a thirty-minute preliminary examination. Without articulating any further deductions, he sent the body off to the medical examiner in Baltimore.

The autopsy was performed the next day, and Dr. Stephen Adams recorded that the deceased was a five-foot-seven-inch, 144-pound white male. He then quickly determined that a single gunshot behind the left ear was the cause of death; it had fractured the skull as devastatingly as if a grenade had exploded inside the cranium. Establishing the identity of the deceased, however, proved more difficult. The corpse's blood type could not be resolved since there was insufficient blood remaining in the body. Fingerprinting was also a challenge since the skin on the hands had decomposed. And a dental identification would be difficult, too, because only partial upper and lower plates remained intact after a week in the bay.

Perplexed, Dr. Adams made the grisly decision to sever the corpse's hands. Packed in ice, the two bloated hands were shipped to the FBI laboratory; the hope was that its technicians might somehow be able to lift a print.

Yet that same day, the chief medical examiner, Dr. Russell Fisher, took it upon himself to ignore his subordinate's lack of conclusive evidence. He did not hesitate to sign a report declaring that the deceased was John Arthur Paisley. The fact that the papers were dated October 1, 1978, and his office had not even received the corpse till the following day attracted no protests. Nor did Dr. Fisher's assertion that the identification was possible thanks to the FBI's fingerprint files, when, in fact, the two disembodied hands had not yet even arrived at the bureau's laboratory. The official cause of death—suicide.

Several days later the body was picked up by a suburban Virginia funeral home that had a cozy relationship with the CIA. It remained stored as if forgotten in the basement for nearly a week until Maryann Paisley, who had somehow never gotten around to viewing the corpse, obligingly signed the order for its cremation. Later, though, she'd assert that she'd never signed *anything.*

With the remains reduced to ashes, it seemed likely that any further interest in the suicide on the Chesapeake Bay would go up in smoke, too, its unresolved mysteries floating off into the ether. The institutional process of forgetting could begin.

BUT THEN, AS FATE AND alcohol would have it, a reporter at a small Wilmington, Delaware, newspaper met up with a Coast Guard buddy for a few beers. In the course of what turned out to be a long boozy evening, the sailor shared a provocative story. The following day the reporter, ignoring his pounding hangover, began poking around. Ten days later he published a very thorough account of "the strange death of a spy."

Once the front-page piece ran, it was the death blow to any hope of keeping the incident quiet. The national press jumped into the fray, and their suspicions, as well as their tenacity, were exacerbated because the CIA, despite its years of practice, remained such poor liars.

Responding to questions, a bemused CIA spokesman publicly dismissed John Paisley as a "low-level analyst." As if lecturing particularly dense young children, he patiently explained that Paisley had retired several years ago, and when he left the agency he'd also, of course, left behind any access to top secret documents. The press, he sternly admonished, was looking for a story where none existed. Move on, he advised.

In swift order, reporters succeeded in shredding his official disclaimers. It was established that John Paisley, in the course of his distinguished twenty-year career, had in fact risen to the upper echelons of the agency, a trusted participant in many of its most consequential operations. And while Paisley had ostensibly retired two years ago, at the time of his death he nevertheless was still very much involved in hush-hush work at the highest levels. He continued to have operational access to the agency's most closely guarded intelligence secrets—its sources and methods for knowing the latest Soviet nuclear developments.

Then, when the CIA's unrepentant spokesman tried to peddle the story that the electronic equipment recovered from the *Brillig* was run-of-the-mill stuff, that, too, turned to sand. A relentless flurry of exposés revealed that the sloop was outfitted with a top secret burst transmitter, that the device worked on a preset frequency to transmit or receive tens of thousands of words per minute, and that the machine was largely used by spy agencies for covert communications with satellites. However, whether the transmitter was one of ours or one of theirs, or, equally incendiary, whether Paisley was communicating with our birds or the Russians, well, that remained the stuff of heated speculation.

With the smell of fresh meat in the air, the newshounds went on the hunt. How did Dr. Fisher, challenged one skeptical journalist, ascertain that the body he'd seen was Paisley's? A close look would've been of no value; the body was unrecognizable, layers of skin eroded by the long emulsion in the bay. All the hair on the corpse's head was gone; same, too, for any trace of Paisley's telltale beard. And as for fingerprints, even if the severed hands proved, against all odds, printable, another problem had, incredibly, materialized: Both the FBI and the CIA claimed they had no sets of fingerprints for John Paisley in their files. And while the CIA spokesman conceded that it was standard practice to fingerprint all agency employees and store the records with the bureau, somehow, he explained as unashamedly as if he were a straight man delivering the punch line to an outrageous joke, Paisley's prints had been "inadvertently destroyed."

More disconcerting findings followed: Paisley's Merchant Marine file listed him at five feet eleven inches and weighing 170 pounds. The autopsy reported that the corpse had apparently shrunk to five feet seven inches and 144 pounds. The remnants of the underwear on the recovered body had a clearly marked 32-inch waist. The BVDs filling a drawer in Paisley's apartment were size 36. And Dr. Weems, the county coroner who was the first professional to gaze at the corpse pulled from the water, gave an explosive interview stating that the neck "looked like it had been irritated, like a kind of squeeze, or there had been a rope around the neck. You get that type of lesion on your neck from hanging." He told reporters that he'd stake his reputation as a coroner, and he'd been in the post for twenty years, that the circle around the body's neck was made before the individual had been killed. Forget the Baltimore medical examiner's suicide verdict, he went on angrily. This was a case of murder. And when diligent journalists gave the autopsy report another perusal, they found that it had dealt with the provocative implications of the neck abrasions by simply not bothering to mention their existence.

In fact, when scrutinized with renewed attention, the official verdict of suicide seemed to fall apart. The gunshot wound was above the left ear. Would Paisley, who was right-handed, have chosen to reach awkwardly across his body to deliver the fatal shot?

And then why the thirty-eight pounds of diving weights? Was that Paisley's attempt to cause his body to sink without a trace so that his suicide would not be revealed? But if so, why? He had two insurance policies, and they both would pay out regardless of the cause of death. And what about the fact that there was no blood, brain tissue, or, for that dumbfounding matter, a gun or simply an expended cartridge aboard the *Brillig*?

The Maryland state police cleverly swept away those annoying details. Paisley, according to the scenario they shared, had criss-crossed his body with the thirty-eight pounds of lead weights, then trundled to the side of the boat and, with one colossal effort, leaped overboard—yet deftly managed to reach across his chest and, while in midair, shoot himself above the left ear. It would have been, some wags suggested, an acrobatic maneuver nearly as nimble as the police investigators' explanation.

Following the unfolding events from afar, exiled to Brussels, retired, privy to no agency files, relying on what he read in the papers and his faint memories of the John Paisley with whom he'd once worked, Pete could only wonder if it would ever be possible to get definitive answers. Was that Paisley's body that had been hauled from the bay? Had he taken his own life? The reality, he decided, was that it would be impossible for him, for any outsider, to make sense of the sort of demons that would drive someone to blow his own brains out. There were things he would never know, Pete thought.

Yet at the same unsettling time Pete grew to feel with an increasing sense of urgency that there were things he needed to know. He thought of Trigon, and the completely unsatisfactory stories offered to explain his unmasking. What had really happened in Moscow? He also found himself wondering, If that wasn't Paisley's body, then whose was it? And why an impostor? Why fake a death? Why, for that matter, christen a boat with a bit of gibberish from Lewis Carroll? Was it, he wondered, some sort of clue? Would only a vorpal blade be able to cut through the thicket of suspicions and half-truths? And he wondered: Was the mystery of Paisley's grotesque death another piece in the long-running puzzle he had been trying to piece together for years? Or was he once again, as his detractors had sneered, crudely shoving square pegs into round holes?

He thought of their vituperative attacks. *Sick think. Paranoid. Obsessed. The Monster Plot.* They had hurled those accusations at him, and, he knew, the imprecations had found their mark. He had been branded. In the end, more angry than beaten, he'd run off, slamming the door firmly behind him.

That should be that, he thought. But he also knew their facile psychologizing had not come close to what had been driving him on back then. Or now. He was his father's son, and the irreducible core of his proud inheritance was an unwavering commitment to his duty. He knew now, as he had always known, that things had gone very wrong at the agency. It was a knowledge that was both cruel and dangerous. And it would not let him rest.

Chapter 3

AND YET PETE HESITATED. BRUSSELS had been in his time a good place to be a spy, and now Pete, nearly eight years after walking away from the agency yet still a sprightly fifty-four, was finding it an even better place to put the past behind him and reenter the overt world.

As the up-and-at-'em 1980s dawned, Brussels remained more a sleepy little old-world village than a city straining to get in on the action. The cobblestoned streets smelled of roasted chestnuts and chocolate. There were wine shops on every corner, and for 100 francs you could buy a bottle of the local Aldeneyck pinot noir and get enough change to purchase a fresh wedge of remedou cheese as an accompaniment. His wife had a convivial circle of friends who were always available for a hand of bridge, and it made Maria quite happy that her family, spread out in Hungary and Austria, was a manageable drive away. As for the kids, well, all three were making their own purposeful ways in the world, and yet they still returned home for the holidays and gathered around the kitchen table for meals that stretched on for hours, thereby keeping him proudly informed about their burgeoning careers.

It had been the mother of his son Andrew's best friend (who, further girding the rightness of their pleasant little world, was also one of Maria's bridge partners) who two years earlier had told them about a soon-to-be-available apartment; and, after only a quick walk-through, they'd made up their minds to move on from their quaint house in the suburbs with its bedrooms under the timber eaves and the cornfield across the road. On a corner of the Avenue de l'Orée, their new home spread out palatially across the entire sun-swept top floor of a redbrick block of flats; a continuous band of windows wrapped around the space, and in the afternoon beams of sunlight sliced across the large rooms like swords. The Bois de la Cambre, an ornate nineteenth-century jewel box of a park, with its emerald-green lawns and a long allée of poplars that led to a wide, cerulean-blue boating lake, was only a stroll from its doorstep. And where the park ended, a vast medieval forest, as dark and as secretive as any woods in a Grimm fairy tale, began. Pete, the amateur (yet quite knowledgeable) ornithologist and dendrologist, would happily trudge over the well-worn trails, ears tuned to the chirp of birdsong, eyes assessing the towering trees.

Yet when he sought refuge, the apartment's study was his chosen hideout. It had originally been part of another flat, and, although it was still necessary to trundle up a few steps to get to its door, it had been connected decades before they'd moved in. A prosperous lawyer had used it as his office in the 1930s and had paneled the walls; when lit by the glow of the fire in the hearth on a frosty night, the old wood took on the sheen of thick, rich honey. A long mahogany dining table had been commandeered to serve as Pete's desk, a cracked leather chair picked up at a flea market stood behind it. In case there was a visitor, a ratty, not very welcoming, once white sofa had been shoved up against an adjacent wall. And dominating the room, reaching from floor to ceiling, were bookcases, the tilting shelves crammed, and yet still more volumes were stacked haphazardly across the floor. Pete read voraciously: European history, Shakespeare, botany. He had many interests, and all were passions.

It was a good life that he and Maria had made. They were very happy.

And Brussels, as well as his expatriate's detachment, had been his choices, his decisions. Back in 1972, after five productive and largely gratifying years as the CIA's station chief in Brussels, it'd been time for a new posting. The bosses wanted him to return to

Langley, to come in from the cold and sit at a desk. But as he mulled that option with Maria, the pot quickly began stirring with other considerations.

For one thing, a headquarters job would require uprooting the children—the two girls were teenagers, their son, nine—and bringing them back to an America that, when looked at from the perspective of their genteel life abroad, seemed a seething cauldron. Drugs, promiscuity, student protests—growing up loomed as a fraught journey in even the most gilded of American schools and suburbs.

For another, a recent change to the federal pension system would allow Pete to add his years in the Marines to the time he'd already put in at the CIA. And when Pete did the math, he'd discovered he could now retire at an energetic forty-six with a full pension. He wouldn't be rich, but the kids wouldn't have to worry about going to college either.

And, more propitious timing, a friend in Brussels had just approached him with a business proposition: purchase surplus materiel from NATO bases—boots, raincoats, whatever—and resell the goodies to other, more deprived corners of the world. Pete, who always enjoyed a new challenge, decided the prospect of becoming an entrepreneur was intriguing.

But in all his discussions with Maria, something else, however unarticulated, had hung over their deliberations. In 1966, and throughout the long, frustrating year that followed, the Counterintelligence Division had turned one of their bloodhounds loose on his professional life, peering into its nooks and crannies. The suspicion that he'd been a traitor, that he'd been the mole who had burrowed deep into the agency, was so patently absurd that Pete had never fretted about the outcome. Of course he'd been cleared, and, further endorsement of his exoneration, then dispatched in late 1967 to Brussels as station chief, no small plum. And yet . . .

There had been a time when, as Richard Helms, a colleague who had ascended to the top job in the agency, recalled, "I believed

Pete would be director of Central Intelligence someday." Pete had it all: a sprawling web of family connections that spread throughout official Washington; a derring-do Cold War fought in the back alleys of Europe; and, no small advantage, he looked every inch the dashing secret agent—tall, broad-shouldered, with an insouciant thatch of straw-blond hair, radiant blue eyes that fixed on you like magnets, and a toothy all-American grin. When he was awarded the prestigious Intelligence Medal in 1965 by the agency director, it was widely believed that soon Pete would be the one bestowing the commendations.

Still, it wasn't just that Pete had been under suspicion, and nasty whispers have a tenacious way of lingering. Rather, as so often happened at Langley, doctrines changed. A new guard had come to power, and one of the new director's first acts was to pin the Intelligence Medal, like the very award that with some ceremony had been given four years before to Pete, on his chief detractor. It was the official signal that Pete's way of looking at the world, his conviction that a long-running Russian plot to infiltrate the agency had succeeded, was out of favor. "Sick think," the Soviet Division now sneered with derision. While Pete, vexed and acerbic, would counter that "the truth had been buried under layers of lies so often repeated that they had become conventional wisdom."

If Pete were to return to an office at Langley, he'd need to discard all his deeply held dogma. If he wanted to be relevant, to continue to be a player in the Great Game, he'd have to move in lockstep with the troops, or else he'd be trampled. Reflecting calmly about his future, he saw his choices: jettison everything he knew was true, even dangerous, or die a slow institutional death. The downward trajectory would be inevitable: He'd be relegated to one smaller office after another, his in-box growing increasingly empty, the up-and-coming young Turks would see him pass in the corridors and snicker at the sad old relic of the Cold War. "The game was over," he conceded, "and we"—his generation, his fellow truth seekers—"had lost."

Better, he'd decided, to pack it all in. While it was still his choice. And if he'd had any doubts about the shrewdness of his decision, about what his institutional fate would have been, he had only to look at how they had sandbagged his friend and sometime mentor Jim Angleton. Angleton, the fabled head of Counterintelligence, had, to Pete's surprise and enduring gratitude, taken him under his wing when Pete had first worked in Washington. And while Pete would be the first to acknowledge that not everything Angleton had done over the years was wise or even proper, he also believed that Angleton was not the mythic villain his enemies had concocted. In fact, in Pete's eyes he was more often right than wrong.

Yet the long knives had come after Angleton. At first the top floor had been content simply to offer hints that it was time to go, that his way of looking at things was counterproductive. When this failed to get results, the attacks became more personal. They whispered about his mental state, sneering that he was unhinged, clinically paranoid. Nevertheless, Angleton had tenaciously hung on. And so finally, in their desperation, they succeeded in tarring Angleton with a judicious leak to the *New York Times*. That Angleton had not initiated the illegal counterintelligence activities that were revealed on the paper's front page did not matter. It was still the perfect excuse to end Angleton's career. "Angleton," the director would write as though jettisoning an underperforming investment in his portfolio, "had, at least in recent years, become more of a liability than an asset."

And how long, Pete wondered, before he, too, would become a "liability"? How long before they'd find an excuse to go after him? The snide, defaming whispers, in fact, had already started. "The tide," he said, "had already turned against me." And once again the truth would be irrelevant. They'd "rewrite history," Pete sourly predicted, to drag him down. "That is what the CIA does best."

Pete had fought to make the world a better place, done his fair share to keep the Red Menace in its place. He couldn't be expected to take on his own agency, too. Besides, who after a lifetime of loyal service doesn't have some regrets, some unfinished business?

Weighing it all in his mind, Pete had chosen to retire. Looking back at the decision, even in the aftermath of the profound concerns that had seized his heart as the news of Trigon's and Paisley's deaths had made its way to him, as old ghosts suddenly reappeared to haunt his thoughts, he reaffirmed the choice he'd made nearly eight years ago. His favorite Shakespeare play was *Henry V*, and he found himself recalling a line: "In peace there's nothing more so becoming a man as modest stillness and humility." In Brussels, in the good life he and Maria had crafted, he had entered his period of "stillness and humility." It would be folly to upend it.

THE WATERLOO BATTLEFIELD WAS NOT more than a twenty-minute ride in his rickety blue Honda Accord from his apartment, and it was a trip Pete made often and always with relish. During his years living in Brussels, he had read a tall stack of histories detailing the events of that 1815 Sunday at Waterloo and had become, as many of those he'd led over the rutted fields and sloping hills could testify, a knowledgeable and compelling tour guide.

As they described the experience, Pete would invariably park in the shadow of the towering Lion's Mound monument that commemorated Wellington's victory and then he'd lead the way on foot. He'd walk at a brisk pace, his head held high as if he were searching for something in the distance, his thick snow-white hair uncovered by a hat, his vanity invariably scorning cold or drizzle.

The route would vary. But there'd usually be a climb to the knoll where Wellington rode high in the saddle to survey the progress of the battle. And he'd lead the way over the ancient fields to follow in the footsteps of the French general Gerard's charge. "March to the sound of the guns," Gerard had extolled, and Pete, while commending his bravery, would also point out that sometimes courage is insufficient; things had not worked out too well for the French that day. Still, Pete made it clear that this was hallowed ground. "The blood-soaked site," he'd proclaim with the reverence of the old Marine that he was, "of forty thousand honorable deaths."

But while history is the epic story of the past, the present is shaped by the unforeseen. And on a gray afternoon on the plains of Waterloo, as a friend visiting from abroad would vividly recall, Pete's own warrior past came to chilling life. In this memory, the presence of a long-haired man who may or may not have been following them provoked first suspicion, then a building sense of danger. A Moscow Center murder team would do it that way, Pete had at last explained evenly. They'd drive you forward toward the triggerman who'd already be in place, waiting, gun cocked and loaded. And with that preamble, Pete abruptly reversed course and headed straight toward the trailing shadow; fixed him with a steely stare; and then, undaunted, strode past him without further incident.

But there were other times when Pete would go alone to Waterloo, using the solitary ramble over the battleground to sort things out, to make sense of the notions at war in his mind. And so it happened that there was no witness to his peregrination on the morning in the icy winter of 1980 when Pete, with the careful introspection of a born case man, pondered the decision he knew he had to make.

Still, it's easy enough to imagine Pete's asking himself—questions he did, in fact, articulate on subsequent occasions—if what he believed was true, if the connection between Trigon and Paisley was indeed a direct one, did he really have a choice? Did the attractions of his comfortable life count for anything when measured against his convictions? For if he was right, and he had little doubt that he was, then the agency, his country, was in genuine danger. With those stony convictions taking hold, who could have blamed Pete if at that decisive moment another line from his favorite Shakespeare roared in his now hardened mind: "Once more unto the breach, dear friends, once more."

And certainly, it is a matter of record that not long after his reflective day at Waterloo, Pete locked himself in his study. With a newfound fluency, he typed out the precepts of his battle plan: a return to action. He would, he wrote with a seemingly feverish intensity, "bring old mysteries out of the dark." He would follow a trail that was "a continuum of treason." He would "set out on my own." The hunter would at last find "the mole who had never been uncovered."

PETE, AT HEART A FAMILY man, waited until the Christmas and New Year holidays had faded into happy memories. Then he booked a ticket to Washington. He did not use an alias; he hadn't thought to cache a passport with a workname. Yet he should have known there was no such thing as a retired spy.

Part II

A Family of Spies

1954–1984

Chapter 4

TO HEAR ANY OF THE Bagleys tell it, one family story that never got old even after all the recounting was the one about Maria's cookbook. Whenever the family gathered around the table for a holiday dinner, someone would start in, and the belly laughs would follow.

It went, more or less, like this: In the days of Vienna's Hapsburg splendor, Maria's family had lived an enviable life cared for by a retinue of attentive servants, including a cook. But the war brought that to a sobering end, and after being uprooted to Hungary, by the time the family had returned to Vienna, about all that remained of their previous grandeur were the cook's handwritten recipes, now bound together with a bit of string. This makeshift cookbook had been passed on to Maria, who, in turn, shared its wisdoms with her daughters.

The recipes were a decidedly archaic remembrance of things long past, and in an age of well-stocked supermarkets, they provoked more mocking laughter than longing. There was, for one cherished example, the instructions for cooking a chicken. "When you pick the chicken," it began helpfully, "the chicken must not be too young or too old." Next, it instructed, hold the struggling bird firmly in place with one hand, while at the same time using your free hand to bring the hatchet down swiftly on the exposed neck. A single blow should do the trick, if executed with speed and precision.

This heirloom cookbook might very well have been on Pete's

mind as he considered the dinner party he was about to attend. It was 1981, the first fresh days of spring in a leafy DC suburb, one of the sleepy Beltway commuter communities to which the nation's espionage establishment returned when their clandestine workday was done. He had been back in the States for several weeks, and for much of the time as he made discreet inquiries, trying too often without success to pry loose the confirming evidence he needed. He had felt an awareness of reliving a former life: the spy's state of perpetual trepidation, the unease that comes from not knowing what lurked around the next corner. Tonight would bring with it the prospect of new opportunities, as well as a tangible reminder of what had pushed him toward retirement in the first place. And while his host, Peter Deriabin, would, he suspected, have no patience with fine cooking, Pete had little doubt he could quite easily snap a chicken's neck or, if it came to it, a man's. After all, you don't get to be a KGB major or Stalin's bodyguard without being made of stern stuff. Nor, for that matter, do you decide to leave behind a lifetime of allegiances and defect to the CIA without some mettle.

And this evening, adding one more role to the many he had assumed in his varied life, Peter Deriabin would be playing Cupid. The targets of his intrigue were two young spies; and the fact that one was Pete's daughter and the other was the son of a power in the Counterintelligence Division who had stood by passively while one of its gung-ho officers from the Special Investigations Group had turned Pete's life inside out—well, that just made Pete's own operational agenda for the evening a bit more thorny.

PETE HAD FIRST MET DERIABIN when the Russian was on the run. On a wintry evening in postwar Vienna—February 15, 1954—when an icy wind blew in from the steppes and gusts of swirling snow camouflaged the scars left by Allied bombers, Deriabin made his move. He had been waiting for the right time since he'd been assigned to the KGB residency, but after a month of reflection, moments of resolve followed by indecision, he had come to realize there was no right time. Not in Vienna.

Vienna in those days was a city where bad things happened, a spoil of war divided by four separate conquerors. Well-armed American, British, French, and Russian soldiers prowled neighborhoods crawling with frenetic black marketeers, desperate refugees, scheming ex-Nazis, and adversarial legions of secret agents. Cross a street, and you could be leaving the American Zone and entering Soviet-controlled territory—and then there'd be no telling what might happen. The Russians had a habit of grabbing anyone that caught their eye; if the suspect couldn't explain himself, the odds were he'd never be seen again. For a spy, Vienna was an adventure, day after lonely day balanced on the knife edge of terror.

Near the center of the city, plunked down like an intruder among the genteel old-world architecture, was the Stiftskaserne. It was an ugly concrete bunker that rose up from the street like a big gray fist. The Nazis had erected it as a brutish symbol of the Reich's intimidating military might; it wound up being used as bomb shelter, Wehrmacht soldiers cowering behind its thick walls from the Allied planes. Now the Americans had laid claim to the structure, making it the stolid centerpiece of their martial presence in the city. It was here that KGB major Deriabin, four-time wounded hero of the battle of Stalingrad, one of just one hundred or so survivors from a regiment that started out with twenty-eight hundred men, Stalin's no-nonsense bodyguard at Tehran and Yalta, and now chief of Counterintelligence for Moscow Center's Austro-German residency, chose as his gateway to a new life.

Deriabin spoke little English, and the few words he struggled to string together were, with sound reason, delivered in a low whisper to the corporal standing guard.

The sentry had no idea what the squat man in the ill-fitting overcoat was mumbling. He gestured dismissively with his rifle, eager to get rid of the nuisance.

"Soviet officer," Deriabin tried, fighting off a rush of panic.

"You want me to call a Soviet officer," the army corporal translated, and his hand reached for the guardhouse phone.

"No! Stop!" Deriabin whispered with taut desperation. "American. American," he corrected urgently.

At last the corporal seemed to understand, and the duty officer was summoned. He listened with some attention, and while he didn't speak a word of Russian, he managed to get the gist: He was staring into the determined eyes of a Soviet intelligence officer who wanted to defect.

It wasn't long before Deriabin was sitting in a warm room with a Russian-speaking CIA officer who identified himself as "Ted." Ted, of course, was a workname, and he treated his guest with a well-honed suspicion; walk-ins were fairly routine occurrences in Vienna. Desperate con men and charlatans making all sorts of claims, bestowing grandiose credentials on themselves, fabricating anti-Communist worldviews as their ticket out of the deprivations and uncertainties of their nasty postwar lives were always, or so it had gotten to seem, knocking on the front door.

The drill, nevertheless, was to hear out the would-be defector. Make no promises, give no assurances—just listen. There was always the possibility, however slight, that this was the real thing: a genuine Moscow Center hood who had decided to switch sides.

Ted went down the standard list of background questions, all the time taking handwritten notes.

As the interrogation continued, Deriabin, with an attentiveness born of a lifetime of practice, observed that Ted was left-handed. And in that same moment, he recalled a report that had circulated throughout the *rezidentura* about a left-handed American intelligence agent. With a sudden, brimming confidence, he interrupted. You wouldn't happen to know a Captain Peterson? he asked slyly.

Abruptly the CIA officer went on high alert: Captain Peterson was the alias he used as case agent for Sergey Feoktistov, a Soviet economics officer, whom the station was running in place.

His small eyes twinkling, his smile broadening, Deriabin continued. "If you should happen to see him, you might mention that his agent Feoktistov is actually working for us."

With that, everything fitted into place. Deriabin had not only revealed a Russian double agent, but he had also firmly established his own valuable credentials. He was the professional intelligence officer he'd claimed to be.

THE EXFILTRATION PLAN FOR "PETER the Great," as Deriabin was now known to the eager cadre of cold warriors who manned the CIA's Vienna station, was chosen because it was, in the end, the least bad of all the other options. The original scheme had been to fly him out of the city and into the safety of the US occupied zone. That had worked without a hitch in the past; there was a small airfield tucked into a corner of Vienna's American sector. But those defectors were not high-priority targets like Deriabin. It was conceivable—probable, in fact—that the enraged Russians would do whatever it would take to knock the light aircraft out of the sky as it flew over their airspace, international incident be damned.

The highways out of Vienna presented a similar concern. There was a checkpoint going into as well as exiting from the Soviet zone, and even the hard guys in the station were reluctant to sign up for a running gun battle with truckloads of heavily armed Russian troops. By default, they settled on the "Mozart Express."

The Mozart Express was what the locals called the American military train that dutifully chugged each day from Vienna to Salzburg in the American zone. The problem, however, was that the train would need to pass through the Soviet occupation zone—and that's when anything might happen. The usual procedure was comfortingly banal: a matter-of-fact exchange of travel documents between the American train commander and the Soviet army officer who ran the show at the border. But the complacent routine might very well be discarded when the KGB realized an invaluable intelligence operative was on the loose. A close inspection of the train and everyone on board could be ordered, and if that were the case, the Americans would not give up their prize without a fight. A cold war could quickly turn very hot, the sparks in Austria igniting a wider conflagration.

It was universally agreed: The scheme bristled with imponderables. But the two CIA officers making the journey with Deriabin had little choice but to soldier their way through and hope for the best. The man in charge was Bill Hood, the chief of Operations of the Vienna Station. Hood was a sturdy, hard-drinking veteran of a lifetime of secret ops, intrepid missions that stretched from his wartime counterintelligence service in X-2 to the spy vs. spy intrigues of present-day Vienna. For his backup, Hood chose a young CIA officer and Marine veteran whom he knew from shared experiences wouldn't turn and run if things threatened to fall apart—Pete Bagley.

The operational plan was, by necessity, makeshift; there was no predicting what might happen. Deriabin was loaded into the cargo car concealed in a large wooden crate stamped, with deliberate imprecision, MACHINERY. A squad of soldiers was in an adjacent car; they were told to be ready, but nothing much beyond that. And Pete and Hood had the passenger cabin to themselves. As the train lurched forward, they played chess to pass the time, but their minds were elsewhere.

The train steamed on uneventfully. Then in the midst of the Soviet zone, brakes screeched with alarm, and the train slowed to a complete halt. The two CIA men exchanged anxious glances, but neither dared to speak. An unnerving quiet filled the passenger cabin. Either Russian soldiers would climb on board, weapons ready, exercising their right to question anyone and inspect any crate passing through Soviet territory, or the train would resume its journey.

The train, as if immobilized, remained locked in place. Pete had a pistol beneath his jacket, and he rested his hand on the weapon, yet at the same time he was careful not to make any sudden movements. All he could do was wait. The Mozart Express would resume its journey, or it wouldn't. And if it didn't, then they would need to make a choice: surrender Deriabin or shoot it out.

Ten minutes passed, but it might as well have been a lifetime in Pete's unsettled world. Then the train began to move slowly forward, until at last it picked up speed.

Two days later, Pete was holed up with Deriabin in a snow-covered chalet high in the Bavarian Alps, the rooftops of a storybook village spread across the valley below. The debriefing lasted several intense days, and both men knew it would be the first of many more to come. But the initial take brought cheers from Langley, and Pete and his prize were driven over icy roads to a military airfield outside Munich. An unheated C-54 took them on a shivery flight, but when the cargo plane landed for refueling on the island of Terceira in the Azores, a welcoming sun was blazing and the pungent fragrance of fields of flowers wafted over the tarmac. For the first time in days, Pete dared to relax.

AND SO IT HAPPENED THAT on a fateful evening twenty-five or so years later, the prized defector, now a much valued analyst for the CIA, had taken on the role of matchmaker. Manipulation is, at the end of the day, second nature to those who live in the secret world.

It's long been forgotten who had first mentioned to Deriabin that two of the adult children of his closest agency colleagues happened to live in the same Arlington, Virginia, garden apartment complex and yet had never met. The fact that Christina Bagley was a CIA trainee, and that Gordon Rocca was an analyst doing hush-hush stuff at the Defense Intelligence Agency, another spook house, would never have been specifically articulated. Nevertheless, with the tacit code of their shared fraternity, a language spoken in meaningful looks and oblique phrases, the gist would have been conveyed. All that Deriabin explicitly knew was that they were two bright, good-looking twentysomething kids who had recently graduated from college and were forging on in their adult lives in "government work." Still, who could have blamed the old KGB hand if he felt he was arranging a match made in espionage heaven?

A genial host, Deriabin sat at the head of a table full of spies. Pete was positioned to his right, but his attention was not fixed on either the defector he'd long ago smuggled into the West or on the two attractive youngsters who seemed to be hitting it off. Rather, the center of his universe that night was a last-minute guest—the Rock, as the old hands at the agency called him.

He had been hastily invited when Deriabin learned his wife was out of town visiting her mother; and as long as the son was coming, the host had figured, why not the father? Yet had Pete been asked, he'd have had no trouble answering; the reproachful memories remained fresh. But instead he'd been taken by surprise. And given his highly charged mood it was now quite possible that the jolly dinner might very well turn into a war party.

Chapter 5

1947

HALF A LIFETIME EARLIER, WHEN Ray Rocca, a skinny young OSS veteran from San Francisco, had teamed up with Jim Angleton, the Rome-based head of American intelligence in Italy during the rough-and-tumble days after the war, he had a reputation as a crack shot with a pistol. It was this skill that had fueled the stories that Rocca had been cast to play the brawn to his boss's brains. The reality, however, always was more complicated. Rocca had a master's in history from Berkeley and was a deep thinker in his own right. And burying the initial misperceptions even deeper, in the course of their intimate partnership any ostensible differences between the two men seemed to fade.

Angleton was, famously, an inscrutable, even mythic spymaster who, after he returned full-time to Washington in 1954, furtively pulled the operational strings in a succession of audacious Cold War intrigues across an embattled globe for the next two decades. His official title was chief of the Clandestine Services Counterintelligence Staff, and his primary mission was to ferret out any moles—the true believers who became eager double agents as well as the traitors who sold out the Republic for thirteen pieces of silver—who had burrowed their way into the agency. He roamed through an opaque terrain he christened "a wilderness of mirrors," and the fact that the master spy had lifted the elegant phrase from T. S. Eliot had only served to embellish his legend as the gentleman spook. Yet at its nasty core, his was dirty work. It was a demanding, quixotically mercurial battlefield, one whose only constant was suspicion. No one was inviolable. The reigning premise was that the agency had been penetrated, and therefore everyone was in his crosshairs, guilty until proven innocent.

Angleton ruled from a spacious second-floor throne room at the agency's headquarters, Venetian blinds drawn, a single lamp illuminating the documents scattered across his oak desk, a cave of shadows. Was this setting an inspired bit of theater, a carefully arranged stage set to reinforce the cloak-and-dagger reputation he was attempting to cultivate? Or was it simply a reflection of his secretive nature? Over time, the motivation grew to be irrelevant. All that mattered was the mystique.

Of course it also helped that he looked the part. Pale as a mortician and just as austere, there was a dour formality to his dress, a repertoire that was partial to somber three-piece suits, and, more often than not, a black homburg topped his full head of pale gray hair. Angleton's face was long and desiccated, as if ravaged by a galloping disease. His shrewd eyes, however, shone, but whether that was caused by the thrill of battle or the effects of his customary three- (and not infrequently four-) martini lunches was anyone's guess. And adding further grist to the folklore were his inherited private fortune, his recondite hobbies, his bohemian intellectualism, and the intensity of his conversation, trains of thought that steamed past his often bewildered audiences to distant horizons. The waggish story had it—and it was one Pete enjoyed circulating—that at the time of the clandestine agency's birthing, when the higher-ups were contemplating the design of an official seal, someone glanced at Angleton, all ghostly taut and conspiratorial, and exclaimed, "I have it! That face!"

And Rocca, in the course of their long partnership, had in many ways become more Angleton than Angleton. His trade, too, was in the reflective arena of counterintelligence. His broad title, bestowed by Angleton, was chief of Research and Analysis, but this largely boiled down to Rocca's keeping his eyes fixed on "the serials." The phrase was Angleton's, and it meant constantly sorting through hundreds of loose threads in hundreds of perplexing espionage cases. The challenge was to discover whether any of these stray strands could be tied to the large ball of suspicions that the two men were continually bouncing off each other.

This meditative, mystery-solving activity was pretty much the essence of counterintelligence. And while Angleton had summed up the discipline with a poet's metaphor, Rocca's definition, while more workmanlike, got closer to its active, constantly probing mentality. "Counterintelligence," he stated, "deals with the activities of other intelligence services in our own country or against our country abroad. In other words, CI is precisely what it says: counter intelligence. The term defines itself." And he might have added, those in the CI ranks are waging perpetual war, always on guard against an enemy who has either sneaked into your house or is planning to do so. In the course of their wrestling with these shared fears and calamitous hypotheses, Angleton and Rocca became a team, a unit that thought and believed as one.

In fact, even when they took a break and climbed out of the trenches, they often found comfort in identical avocations. It was Angleton who spent his weekends dotingly cultivating orchids, even succeeding in breeding a prizewinning species he named after his much-aggrieved wife, Cicely. What had attracted him to orchids, he explained, were their deceptive qualities—how the flowers use trickery to dupe insects into flocking to them. And while he never deigned to explain the interconnection explicitly, chicanery was also how intelligence services worked; spy agencies, too, used guile to send their adversaries down dead-end trails or lure them into disasters.

Rocca, though, grasped the link, and he became another orchid aficionado. Two giant cattleya orchids stood like sentinels outside the study of his home in suburban Virginia, and he frequented the same orchid supplier in Kensington, Maryland, as Angleton. And again like his boss (or, as many agency colleagues would affirm, coconspirator), Rocca had taken on the ascetic countenance of a monk who spent his days and nights laboriously poring over forgotten texts. Time had transformed Rocca into a gray-haired, stoop-shouldered presence with a fittingly professorial goatee. Any connection to the sharp-shooting OSS fieldman seemed a distant, almost irreconcilable, memory.

Chapter 6

1966

YET WHAT FORCEFULLY LINKED ANGLETON and Rocca together in Pete's resentful mind was that they both had stood impassively by as their underlings, the Special Investigative Group (SIG) mole hunters, had for one disquieting year made his professional life a living hell. And adding further irony to the dinner party this evening, it was their mutual host, Peter Deriabin, who had figured prominently in a portion of that wildly misguided, egregiously damning, and still infuriating indictment.

THE CODE NAME WAS HONETOL. Was this a random series of letters as was the standard agency practice? Or, as the office scuttlebutt had it, was the name one of Angleton's esoteric acronyms, part of a private code only he could decipher? In the end, though, its meaning became clear: It was a manhunt.

Clare Edward Petty was one of the hunters. He had grown up in Oklahoma, a country boy with freckles and, even as a youngster, an adult's determined, lantern jaw. After the war, he had joined the CIA, and while stationed in Germany he'd identified a double agent in the West German Intelligence Agency (BND). That caught Angleton's eye, and in 1966 Petty was brought back to headquarters as part of the SIG team.

Petty never harbored any doubts about the validity of his mission.

He was a man of deep loyalties, and once pledged they were unshakable. In Petty's judgment, "Angleton undoubtedly understood counterintelligence better than any man alive." The great man himself had set him loose on this hunt, and that was all the justification he required.

On an operational level, Petty's detective work was guided by several assumptions, articles of faith that Angleton and Rocca had instilled. The foundational tenet was that the agency had been penetrated; after all, since the KGB, the headline-making evidence had made demonstrably clear, had succeeded in infiltrating their agents into the upper echelons of the British and French intelligence services, why wouldn't they have had similar success with America's? And, in a corollary insight, the practical Moscow Center strategists wouldn't waste their energies on a capricious low-level penetration. They'd go after a glittering prize, zeroing in on where they could cause the most mischief.

Moscow Center's grand ambition would be to recruit one of the spymasters serving at the pinnacle of the clandestine service, ideally the director of the CIA, or one of his knowledgeable deputies. But even Angleton's agitated imagination (at least as shared with his acolytes) didn't hold that as a realistic danger; exhaustive vetting, he conceded, would already have put the high-level appointees under a microscope. The SIG team therefore focused on a more viable potential KGB target a bit lower in the agency hierarchy—the Soviet Bloc Division.

To the enemy, the SB would make perfect sense. For one thing, the officers in the division possessed precisely the sort of invaluable secrets Moscow Center was eager to get: They'd have direct knowledge of the ops aimed against Mother Russia. And for another, since the division was always rolling out plans to recruit Soviet intelligence officers, it seemed an elementary deduction to conclude that the KGB would be up to similar tricks.

Now that he had narrowed the target, Petty went on the attack. From his office window, Petty would often find himself staring across the courtyard at the second-floor window directly opposite his own. If the blinds on room 2C43 were drawn, that meant Angleton was at his desk. In Petty's mind's eye, he'd envision the master spy focusing his attention on the most secret of agency files, searching for the connections other, lesser minds had missed. This encouraged Petty, goading him on as he slowly proceeded with his own painstaking investigations.

He started with the defector files, the records a long, time-consuming journey backward into some of the SB's greatest triumphs. Yet he had the inkling that something was not right, that some of the celebrations were false fronts. Informed by his new perspective, the history grew increasingly ominous. So he pushed on. He pulled back the overgrown weeds, and in the process, he believed, he'd exposed cracks in the foundation that had been previously ignored. And at last, there it was: The identity of the traitor was within his grasp.

Chapter 7

IN PETTY'S HARD-CHARGING WORLD, THE game was now afoot. The key to unlocking all his suspicions, he'd come to realize, had been buried deep in the many volumes that made up the Peter Deriabin case file. It had been there all along, only no one had noticed. Yet the devil is always in the details. And, as some professionals will also tell you, so is the mole.

Returning to this well-trodden ground, Petty soon found the first intimations of the elusive secret he'd been pursuing. In the early days after the Mozart Express exfiltration, when Deriabin had been holed up with Bagley in the Bavarian safe house, the Russian had been asked to provide the names of the KGB officers in the Vienna rezidentura. He went to work, and the roll call of enemy agents he compiled was duly ascertained as solid-gold intel. Langley offered up official expressions of praise, and claps on the back for Bagley as well as his all-star defector.

But when Petty now reread the laudatory report on this small incident, it provoked a vague yet unsettling consternation. Nothing specific, nevertheless the sense that he was missing something would not leave. And in the restless period that followed, his inchoate instinct began to take a more discernible shape.

Coaxed on by his prodigious memory, Petty at last recalled something he'd read in another defector's interview. A KGB officer had told the CIA interrogators that his assignment to Vienna had been abruptly canceled by his superiors because Deriabin had revealed his name and position. At the time, this intel was seen as simply confirmation of the authenticity of Deriabin's initial disclosures. Only now, from the depths of a mind already swirling with suspicions, Petty asked himself the question he decided the agency should have asked years ago: How did Moscow Center know what Deriabin had disclosed?

With newfound urgency, he hunted down the original debriefing. And this time when he went back over the file, his attention was fixed on something that had been ignored for years: The defector had matter-of-factly stated that his KGB boss had a transcript of Deriabin's interview with Bagley on his desk.

Which meant: Everything he had feared was true. There was a leak in the SB Division. And maybe he had found a suspect.

THE ARCHIVES OF THE AGENCY'S Polish Branch were contained in two dark, uninviting basement rooms. Sturdy cardboard boxes aligned in neat, orderly rows filled metal shelves from floor to ceiling. Inside the boxes were countless buff files, business-like activity reports that contained long-forgotten and, all too often, deliberately buried secrets.

But for Petty, who now had the scent, this was fertile ground. He had a theory, and he needed to test it. He was moving forward driven by little more than a whiff of a memory, but it was sufficient to lead him to where he felt he first needed to go. He pulled down a heavy box. Inside was the case history of Michal Goleniewski, the Polish intelligence officer who had christened himself "Sniper."

The tale he read began like a pulp spy novel. Out of the blue, in April 1959, a letter addressed to "the American ambassador" arrived at the embassy in Bern, Switzerland. The writer announced that he was a Soviet intelligence officer from "behind the Iron Curtain." Typed in a literate German, it didn't offer much that was promising, certainly nothing that the CIA's Bern Station hadn't already known. But, tacitly acknowledging that he'd delivered peanuts, the writer provocatively offered that in subsequent letters he could provide "more interesting" material. He signed the letter "Heckenschütze," which translated as "sniper." But whether he was taking aim at the Russians or at the Americans quickly became a matter of hot debate. And Pete Bagley, read Petty as he followed the paper trail of reports with mounting interest, was smack in the middle of the ruckus.

At this point in his high-flying career, Bagley, after serving a tour back in the States in the Polish Division, had recently been appointed second secretary at the embassy in Bern. The diplomatic posting was cover—and a transparently thin one at best—for his busy spy work; staid, prim Switzerland was the springboard for the enemy's machinations beyond the Alps and across Eastern Europe. As part of the elite Soviet Division, Bagley's job was to focus on REDTOPS, which was agency speak for Russian diplomats, military bigwigs, and intelligence officers temporarily plying their sinister trades in the West. And by "focus," headquarters meant "recruit." In due course, the Sniper letter fell into his hands.

It was Bagley, Petty discovered, who had persuaded the station's naysayers to cast aside their fears of a disinformation op and play ball with Sniper. An advertisement, as the anonymous writer had requested, was then placed in the personals column of *Frankfurter Allgemeine Zeitung*; it included an address where subsequent letters could be sent.

Sniper proved to be a diligent correspondent. And the lengthy, single-spaced letters he sent were pure intelligence gold; "invaluable," Bagley officially gloated, now certain any early concerns had been unfounded.

But Petty, looking at the case with the infallible accuracy of hindsight, noticed that there had come a time when the secrets Sniper shared were not just intriguing—they were *precisely* the sort of information the agency had been busily trying to dig up. And Petty had been a professional for too long to believe in fortuitous coincidences. He reasoned that the KGB had stumbled onto the game Sniper was playing and had begun manipulating events to have the last cynical laugh. "They were using him for aggressive CI," Petty deduced. Without Sniper's knowledge, they were skillfully feeding the traitor "secrets" they wanted him to share with the West.

Which left Petty asking himself: How did the Russians stumble onto Sniper?

It didn't take the mole hunter long to verify that there were only four officers in the SB Division who were aware of the Sniper operation. And Pete Bagley was one of them. Which proved nothing. But it got Petty thinking.

Given Petty's predisposed mind, his suspicions quickly crossed the Rubicon. He was convinced he had identified the traitor. And, plowing on through the files, he found the confirmation he needed.

In the midst of the first heady days of the Sniper operation, Bagley had been running an op to recruit a vulnerable UB (Polish Intelligence Service) officer. The targeted agent had played fast and loose with his service's money, and Bagley, waving the false flag of a West German intelligence officer, offered to come to his rescue. A trade was suggested: the money the agent needed in exchange for a piece of his soul.

The hard-pressed Polish spy indicated that he was amenable to any sort of arrangement that would spare him from his vengeful colleagues. So a date was set to seal the deal. But when Bagley showed up, the spy abruptly turned him down.

No buy, Bagley signaled headquarters. As for an explanation, he didn't offer much more than a philosophical shrug. Burning an agent is a tricky business; you just never know if he'll get a sudden dose of religion and decide to face up to the consequences rather than turn traitor.

But elsewhere in the compendious files, Petty located a report that offered a more concrete rationale for the Polish spy's sudden change of heart. An asset the agency was running in the UB had reported that two weeks before Bagley's planned meet with his Polish target, headquarters had received a Flash cable from their bosses in Moscow Center: A West German intelligence agent is preparing to approach one of your officers. The would-be traitor's name was provided, as well as the stern instruction to nip the scheme in the bud. Which, of course, was precisely what had occurred.

Now, once again, Petty couldn't help but ask, How did the KGB know? They had all the details of the approach, down to the West German false flag that had been blowing in the wind. There was one man who certainly would've been in a position to disclose that closely guarded information. When Petty did his sums, he had his answer. The evidence was sparse and woefully circumstantial to boot, but in Petty's roiling mind suspicion had overwhelmed hard facts. In his eagerness to succeed, he was ready to jump high and far to predetermined conclusions.

"Pete was a good friend," Petty would say. "But in the end you eliminate the possibilities." The mole hunter wrote up a lengthy report that concluded that "Bagley was a candidate to whom we should pay serious attention."

Emboldened by a sense of duty, yet at the same time aware of Bagley's golden boy status in the agency, he submitted his analysis—Bagley was referred to in the pages as GIRAFFE, his cryptonym—to James Ramsay Hunt, Petty's immediate superior. At that moment, Petty was quaking with trepidation, prepared to be chastised like a schoolboy for his audacity.

"This is the best thing I've seen yet," Hunt congratulated Petty after he'd finished reading the indictment. He would immediately pass on the report, he informed Petty, to Angleton.

Weeks passed, but Angleton remained silent. Petty could only wonder if he was going to get a medal or shown the door. He had no idea what to expect. After all, Bagley and Angleton were famously close. Yet Petty reminded himself he had had no choice. His commitment was to the agency, not to his career, and certainly not to any individuals, whether old colleagues or even mentors.

At last, Angleton summoned Petty to his lair. He talked about extraneous matters, and Petty found much of the lecture inaccessible. Finally, as if it were a desultory aside to their more pertinent conversation, he addressed Petty's report on GIRAFFE.

"It's not Pete," Angleton declared with an acid finality.

PETTY WAS LEFT REELING. His long solitary undertaking, years of exhaustive labor, had been dismissed without any comment other than a three-word verdict. Abruptly, he had been sent scurrying back to square one.

And if Bagley wasn't the mole, then who was? Where do I go from here? Petty wondered, unable to suppress a peal of self-pity. His mood seesawed dangerously; one moment he was wallowing in his defeat, the next he wanted to scream with rage.

Yet at the very time when he was struggling to find the will to resume his mission, a sudden sense of recognition obliterated all his previous frustration. It was a eureka moment, and it came to him in a flash.

"I decided I'd been looking at it all wrong . . . I began rethinking everything," he would later explain. "If you turned the flip side, it all made sense."

"The only place in the CIA besides the cable room where there was total access was on James Angleton's desk. From the indications we had, the penetration had to be at a high and sensitive level, and long-term. You could say the director's desk fit that description, but there were several directors. All the operational cables, however, went through Angleton."

Spurred on by these wild deductive leaps, Petty declared he had at last managed to split the atom: "Angleton was the mole."

And just like that the spring was once more in Petty's stride. He went back over years of covert history with renewed passion, and despite all the time he'd already put in, he refused to be rushed. He was a lone avenger, but he didn't mind; let the rest of the SIG be damned. He was driven by the burning fervor of someone who knew he was right.

He shared the existence of his secret mission with only a single agency official, James Critchfield, his mentor during Petty's dangerous days in Germany. For two years he doggedly toiled away, a spy spying on the spymaster, the underling covertly collecting what he felt was a mountain of damning evidence against his own boss.

When, by the spring of 1973, he'd come pretty much to the end of the line, Petty made a quixotic decision: He'd retire. He had had his fill of an agency where a mole—and by rotten extension the enemy—was turning the place inside out. But before he walked out the door, he needed to speak to someone. He wanted to share what he'd learned with William Colby, the new director. The closest he could get to the pinnacle, however, was David Blee, the deputy director of Operations.

Petty told him everything. The words poured out in a torrent. He held nothing back, however circumstantial. It was a purging, cathartic exercise.

Blee listened, rarely interrupting. His silence was inscrutable.

When Petty finished, Blee politely thanked him for the presentation; and then, without further significant comment, announced he had a meeting he needed to attend.

A week or so later, Petty was summoned to Blee's office. "We'd like you to stay," he announced. By "we," Petty understood that he meant Director Colby. "The clear implication," Petty recalled, enjoying what he felt was his vindication, "was 'Keep doing what you're doing.' So I stayed and kept at it."

But after another grueling year, Petty had finally had enough. He had started to feel that if he kept at it, his efforts would ultimately prove as destructive as the mole he had been pursuing. His hunt would also pull the agency apart. Once again he made up his mind to retire, and this time the decision was firm.

Before he left, Petty sat down with James Burke, a senior officer. They talked over the course of four days, a total of twenty-six hours, the tape running all the while. In addition, Petty turned over two safes full of material that, he confidentially declared, were a road map that would show the way to confirming all his accusations.

When he was done presenting his case, Petty didn't hesitate to state clearly his bottom line. He said Angleton was the mole, and he should be fired. As a final parting shot, he added that Angleton's three deputies—Rocca and two others—should also be dismissed. You had to assume, he said, that they could be controlled agents.

In the days that followed, Angleton wasn't summarily fired. Instead, Colby, always the sly bureaucrat, humiliated him, stripping away one long-held institutional power after another. "It was designed to lead him to see the handwriting on the wall," conceded Colby, who, even before being briefed on Petty's sensational accusations, had had as much as he could stomach of the agency's vindictive mole hunts. "He just wouldn't take the bait."

So Colby waited. And when a damning *New York Times* investigation put Angleton, however erroneously, at the controlling center of the agency domestic abuses of power, he quickly grabbed the opportunity. Colby called Angleton and with a resigned anger said, "You and I know we've had these discussions over a period of time. But I insist you go now."

Later that day, Angleton announced to Rocca that he was retiring, and that Rocca would be reassigned. The director no longer wanted Rocca to work in counterintelligence. "What are you going to do?" he asked.

"I guess I'll retire," said a defeated Rocca.

"Good," agreed Angleton.

In the explosive aftermath, as both applauding and disapproving agency factions had their say, Colby felt it was necessary to add an explanatory footnote to the savage events. "I can absolutely say for certain," he declared, "that I did not fire Angleton because he was a Soviet agent."

And everyone at the agency absolutely believed him. Except for those who absolutely did not.

BACK IN THE PRESENT AT the dinner party, no longer wandering through the memories of old battles, Pete found it was difficult to hold a grudge against the Rock. No doubt a less equable man might have taunted, poetic justice. Your underlings fed on me, and then, still hungry, they took a big bite out of your reputation, too.

But Pete, with an objectivity honed by decades of calm reflection, knew that the Rock was more right than wrong. They both believed the agency had been infiltrated. That fundamental intellectual kinship made it impossible to see the Rock as anything but another old spy who'd fallen from grace. Another cold warrior whose time had passed. Another victim, like himself. And when you came down to it, they had fought many of the same battles, been on the same side in their day. If not friends, well, they had certainly been colleagues. Besides, they both had deep, indelible ties to Angleton.

And so, while the young couple continued their happy chatter, Pete, as had been his intention from the moment he'd sat down at Deriabin's table, steered the conversation to the seemingly unrelated history of an old, inactive case.

Nosenko, Yuri Ivanovich. Lieutenant colonel, KGB Second Chief Directorate, he might very well have begun, as if reciting from memory the heading on the yellowed and voluminous case file.

And while it was old, Pete was convinced it wasn't at all dead. It was tied to everything that followed.

Chapter 8

1962

THERE ARE MOMENTS WHEN, JUST like that, a life irrevocably changes. Even if you're unaware of it at the time, the consequences are inescapable. For Pete, there would always be Before Nosenko and After Nosenko. It would become the defining incident of his career, the motivations for all his dark thoughts, for his sense of terror that did not diminish with the passing decades. It was what had way back in 1962 first presented him with the reality of the hovering danger. And now, twenty long years later, with the certainty he needed to continue to pursue his hunt.

True, he had been pulled back in by the troubling questions—and the two confounding deaths—at the core of the more recent Trigon and Paisley cases. Yet "unraveling knot by knot these twisted strands," he was convinced, "required a long look backward." A lifetime in the secret world had taught him "the past pervades counterintelligence work."

"Nosenko is the Rosetta stone," Pete knew. By that he meant that the deciphering of this long-running case held the key to understanding *everything.* It would reveal generations of the enemy's aggressive schemes. It could even point the way to the traitor—or was it, he often wondered, traitors?—within the agency.

It had dawned on him as he had sat months ago in the wood-paneled serenity of his den in Brussels that this was where his quest must start. And just as Jean-François Champollion, a determined young Frenchman, had kept at it until he'd cracked the Rosetta stone's hieroglyphics, Pete remained convinced that he, certainly no less driven, must make sense of the Nosenko affair. In the autumn of his life, it still spurred him on. And when he succeeded, he'd find, as he put it with an uncharacteristic poetic flourish, "the undiscovered traitors still hovering like ghosts."

On operational footing since he'd arrived in America, these ghosts had been actively spooking his thoughts. Tonight the ex–secret agent, taking advantage of the cover afforded by his paternal role in facilitating the blind date for the two young spies, had come hoping to recruit his wily host's help in digging into their shared haunted past; if the Rock wanted to join in the hunt, too, well, so much the better. And ever since he'd walked in the door, Pete had been on his way back to a bright May afternoon in 1962, a glorious spring day in Geneva's Old Town, when riots of red flowers bloomed in the window boxes, a relentless Swiss sun bore down on the faded wood-shingled rooftops, and his life had changed forever.

EVERY DEFECTOR'S STORY CAN BE said to begin the same way: when the first step is taken to cross over to the other side. For Yuri Nosenko, it began in a long, high-ceilinged marble corridor outside a stuffy conference room at the Palais des Nations. A much-needed break had just interrupted the morning session of a tedious and unproductive arms control conference.

As the liberated delegates wandered down the stately hallway, Nosenko slipped out of the crowd and sought out an American attaché he knew had served in Moscow. The American stood alone, silhouetted by the diffused glow coming from a small window that offered a glimpse of Lake Geneva. Like good diplomats, they shook hands and exchanged hail-fellow greetings; and then, after a furtive glance to assure that no members of the Soviet delegation hovered nearby, Nosenko made his pitch. I need to contact the CIA, he blurted. It's urgent.

The next two days were an interminable lifetime for Nosenko. At last, though, the American diplomat reappeared and led him to the door of a cozy apartment tucked into a block of flats in a lackluster section of Geneva. He quickly introduced the Russian to Pete: "This is Mr. Nosenko of the Soviet delegation." And then, to his apparent great relief, he made a hasty and very undiplomatic getaway.

Pete, who of course was using a workname, had arrived at the safe house from Bern earlier that morning. He'd gotten there in time to make sure that there was a bowl of salted nuts on the table and the bar was stocked; there is a certain etiquette to safe house hospitality. And he'd also determined the concealed tape recorders were up and running. In matters of tradecraft, Pete was always diligent.

"I heard you wanted to talk to someone in American intelligence," Pete, according to the notes he'd later transcribe, began in English. At the same time, he took a quick measure of the Russian. He was big, broad, and yet puppyish; it was as if he were embarrassed to be there, at least that was what the silly grin on his face and hunched shoulders seemed to be signaling. As for age, Pete guessed in his mid-thirties. He had a wide Slavic face, a fighter's flattened nose, and light brown hair swept straight back to expose a mile of forehead. In a crowd, you'd pick him out as the Russian every time. His hooded brown eyes were another tell. They darted about, checking out the apartment. When he took careful measure of the view beyond the narrow balcony, Pete decided he was in the presence of another professional. That's what he'd do, too: make sure you're higher than the adjacent buildings; that way there's no worry about someone filming from across the way.

"I'm pleased to meet you," Pete continued. The drill was to keep things friendly, but not too friendly. Eagerness was never helpful, especially when you don't know what the defector wants, or, for that matter, what he has to offer.

Nosenko smiled at him. And Pete smiled back. They might have been two poker players deciding whether to raise the stakes.

"I have important things to tell you," Nosenko said finally in Russian.

Abruptly Pete raised his hand like a traffic cop. "I understand Russian but have trouble expressing myself clearly in it. Let's speak English."

"No problem," Nosenko said agreeably, this time in English.

Pete pointed his guest to an upholstered chair and, the congenial host, asked if he'd like a drink.

"Yes, please, scotch."

That's the Russian way, Pete knew. Vodka at home, scotch when abroad. He poured the whiskey over a glass full of ice and added a generous dash of soda; the last thing he needed was a drunk Russian.

Nosenko took a large swallow of his drink, and then made his pitch. "I'm in trouble," he said with convincing despair, "I need money urgently."

Like a stony loan officer at a bank, Pete nodded blankly.

"I think you'll help me," responded Nosenko, undeterred. "I am an officer of the KGB, and I work against your people in Moscow."

Hearing this, Pete contrived to appear very bored. But all his internal alarms had started clanging. Landing an active KGB agent would be like winning the lottery.

"I am a major in the Second Chief Directorate," Nosenko went on. Pete's boredom visibly deepened, yet he knew, as would any REDTOP officer, that this directorate handled counterintelligence.

His host's inward silence seemed to nudge Nosenko to continue. "I am responsible for the security of our delegation."

Before he'd left Bern, Pete had done his homework, and he swiftly worked out the particulars. Foreign Minister Andrei Gromyko had arrived from Moscow back in March leading an entourage of seventy delegates, and while Gromyko had wisely hightailed it out of Geneva after the opening session, the others had stayed for the long haul. Nosenko's name appeared on the official roster, but the fact that he was KGB was a bit of news. Still it was, Pete knew, standard for Moscow Center to send a few of their own along to keep a watchful eye; the Kremlin was always nervous that one of their foreign service officers would be seduced by the gaudy delights of the West and do a runner.

Having shared this small secret, Nosenko raised his glass imploringly, as though seeking a reward.

Pete obligingly refilled it, only this time when he went to add a measure of soda, Nosenko brusquely grabbed the glass from his host's hand.

A large swallow, and Nosenko picked up his story. "I need money right now. I've been in too many bars, been with too many girls, drunk too much whiskey." He flicked an index finger against his neck, the way Russians do when they know the walls are closing in on them.

Pete decided he'd better move quickly. You never know when your prize will get up and head for the door.

"How much do you owe?" Pete asked.

"Eight hundred francs."

Pete nodded consolingly, but the math he did in his head was cheering. The amount, although a week's pay for the Russian, was only $250.

Nosenko plowed on: "I'll answer your questions, but you must understand that I will never come over to your side to live in the West. I wouldn't leave my family or my country. I have two little girls."

He proudly shared photos of the children with the American. While all the time Pete was excitedly thinking, This is getting better and better. Hell, we don't want you to defect. We want you to keep working in Moscow Center—and to keep reporting to us.

But today, Pete worried, the clock was ticking away. "I'd like you to tell me," he abruptly demanded, "what *you* think is the most important thing you have to tell us."

Nosenko paused thoughtfully. Then: "I know the most important American spy the KGB ever recruited in Moscow."

Pete did not even try to conceal his sudden eagerness; he knew he wouldn't be able to find the art. He leaned forward in his chair, his face now mere inches from the Russian. He didn't want to miss a single word.

The essence of Nosenko's disclosure was a revelation: An army sergeant working in the embassy as a cipher machine mechanic had plunged with amorous abandon into a KGB honey trap. There were reels of racy movies of the sergeant cavorting with a Russian woman working in the embassy apartments, and he was threatened that there'd be a special screening for his wife unless he cooperated. He was a "tremendously valuable source," so good that Nosenko's boss went to the United States to reactivate him after the sergeant had been deployed home at the end of his tour. As for the sergeant's identity, all Nosenko could offer was a code name—Andrey.

And then all of a sudden Nosenko was in a hurry to leave. "I should go now. But I come back the day after tomorrow." He suggested that sometime in the late afternoon would be best; that's when it was easier to slip away unnoticed from the delegation.

Pete assured Nosenko he'd have the 800 francs waiting for him.

The two men headed to the door. Only as Pete went to open it, Nosenko stopped, looked him directly in the eye, and blurted, "I know how Popov was caught."

If a single declaration had been designed to focus Pete's attention, it would have been difficult to imagine a more successful ploy. The impact was almost physical. Lieutenant Colonel Pyotr Semyonovich Popov had been *his* case. *His* great success. And *his* grievous loss. Pete at one point had been one of the handlers of this GRU (Soviet military intelligence) officer who, for seven daring years, had delivered a brimming treasure chest of political and military secrets to the agency. Then in 1959, Popov had been suddenly arrested; his execution soon followed. It was a colossal intelligence loss for the agency, and for Pete the cruel death of a friend stung deep.

"Tell me how," Pete demanded.

But Nosenko shook his head, and at the same time stepped backward in retreat. "No, no. I don't have the time," he insisted with newly found resolve.

Pete was baffled. A moment before, Nosenko had been chatty. Now he was looking anxiously at his watch. His mercurial behavior made no sense—unless it was a deliberate tease.

"It won't take another minute," Pete implored, realizing even as he spoke how desperate he sounded.

Only Nosenko had already opened the door. A quick peek into the hallway, a whispered "Next time," and he vanished down the pitch-dark stairwell.

"Damn!" Pete exploded as soon as he shut the door. This was not the first time he'd sent a Joe off wondering if he'd ever see him again, but never before had the rupture been so gut-wrenching.

Chapter 9

1962

But Nosenko was Pete's destiny. He returned. And when he did, Pete was ready.

Taking care to use the word code heading that would limit distribution, Pete had sent a Flash cable to Langley within an hour of Nosenko's departure. He reported that Yuri Nosenko was precisely what he'd claimed to be: a genuine Moscow Center–trained professional. Only a KGB operative would have access to the sort of insider's knowledge of past and present ops, secrets still closely guarded, that Nosenko had fluently shared. And he concluded with a request: He wanted help. Send an officer fluent in Russian. The conversation had so far been largely in English, but if in the future Nosenko drifted into his native tongue, Pete wanted to be sure that neither a single word nor a provocative nuance would be lost. The potential yield was that promising, he predicted with absolute confidence.

For once headquarters responded with haste, and only several hours later Pete was reading the decoded reply. There was no record of a Yuri Nosenko in the central files except for someone with that name who had been part of a delegation to Cuba. A cursory search of the defector debriefs had also drawn a blank: no mention of a KGB officer with that name.

The only Nosenko who had turned up in the trace was an Ivan Nosenko, the Soviet minister of shipbuilding and, no less significantly, a member of the Central Committee of the Communist Party. When he'd died six years earlier, he'd been eulogized as a Hero of the Soviet Union. Khrushchev himself had been a guard of honor at the funeral and his remains had been interred in the Kremlin Wall. If the Nosenko who'd walked into the Geneva safe house was a blood relative, the cable suggested, not even trying to disguise the excitement that had been kindled in the SB Division, there might be an opportunity to dig into an untapped mother lode. And as further confirmation of their enthusiasm, Langley announced that a Russian-speaking officer would be flying out on the next available plane.

George Kisevalter, hastily packed suitcase in hand, arrived the next day, and for Pete it wasn't soon enough. Born in St. Petersburg, Kisevalter had worked in tandem with Pete out of the Vienna Station to handle Popov. In the course of this demanding operation, Pete had grown to admire his burly colleague's idiomatic command of Russian as well as his instinctive knack for winning over his operatives. A friendly face and a reassuring pat on the back went a long way to reminding an agent working behind enemy lines why he was taking such risks, and Kisevalter had this indispensable agent runner's gift.

The two officers had decided to hunker down in the cramped safe house apartment. There was no telling when Nosenko would appear, and they wanted to be sure to be there to greet him. They spent a long day playing chess, trading headquarters gossip, and anxiously waiting. At night they flipped a coin for the one bed, and Pete lost and had to make do on the narrow couch. The next afternoon there was a knock on the door.

THE SEVEN DAYS THAT FOLLOWED were an education. There were four separate meetings during that incredible week, and while one was hasty, since Nosenko had to dash off to a conference cocktail reception early that evening, the others went on for three hours or so each. The scotch flowed, too, and that helped to keep the mood loose. It might have been a convivial trio of old-timers trading war stories. Of course, the fact that the one doing most of the talking was a KGB officer and he was reeling off state secrets did add to the drama.

And the secrets that Nosenko disclosed! With no hint of his previous hesitancy, he succinctly explained how Popov had been blown. "Our guys were routinely tailing George Winters, an attaché at your embassy," he said, bursting out into a prideful grin. "They saw him drop a letter into a street mailbox. It was written in Russian with a false return address and addressed to Popov. That was all we needed—diplomats don't post innocent letters to GRU officers."

He also bragged about the KGB's "first-class" surveillance techniques. "There's a powder we call 'Metka' that's put into the pockets of American diplomats. It leaves a chemical trace on any envelope they'd carry for posting on the street. Censorship picks up the trace." Then there was a clear liquid that they'd paint on car roofs; it allowed watchers perched on rooftops across Moscow to track a vehicle with ease despite even the most skillful drivers' attempts to evade detection. And there was "Neptune 80," which, when applied to a target's shoes, left behind a scent that drove the watchers' dogs wild and made tracking a breeze.

As for the foreign embassies in Moscow, Nosenko said that the KGB might as well have had keys to the front doors. "We have great teams," he said, seemingly unmindful that he was betraying the organization he was boasting about. "They know how to get in, open locked safes, take the stuff out and photograph it on the spot and put it back without one thing showing that they'd ever been there."

The sly teams had also spiked the American embassies. He didn't know where the microphones were hidden, but he went on helpfully, "I've read transcripts of conversations in maybe ten different offices."

The real prize, though, was Nosenko's unveiling the identities of several agents Moscow Center had been running in the West. There was a French businessman who was in it for the money. A French ambassador who was being blackmailed after a spirited run-in with a well-trained KGB seductress. A staffer in the British Admiralty who'd been delivering naval secrets for the past five years rather than risk having his homosexuality revealed to his superiors. ("Homosexuals," Nosenko had sneered with a nasty dismissiveness. "We have a bunch of them working for us, ready for jobs like this.") And further demonstrating to Pete that he was the real thing, not just a plant sinking his hooks into the CIA, Nosenko revealed an active double. He disclosed that a Soviet radio journalist who was on the agency payroll had been working all along for the KGB.

In the rambling course of these wide-ranging debriefs, Nosenko not only confirmed Langley's wishful hope that he might be related to the distinguished Nosenko who had been a revered pillar of the Communist Party, but he offhandedly revealed that he was Ivan's oldest son. That would get the attention of headquarters, Pete silently rejoiced. Nosenko, however, seemed more eager to talk about his own family, his wife (his second) and his two daughters. And so when he fretted out loud about his daughter Oksana's asthma and the impossibility of getting a specific medication in Russia, Pete, the shrewd agent runner, jumped into action. A Flash cable to headquarters, and the very next morning the much-needed medicine arrived in Geneva in the diplomatic pouch. Affecting a nothing-to-it casualness, Pete handed the elixir to a grateful Nosenko that afternoon.

The heady days, however, were quickly counting down. The arms control conference was scheduled to adjourn at the end of the week, and Nosenko would return to Moscow with the delegation. At their penultimate session in the safe house, Pete, having played the attentive suitor for the past six days, finally popped the question. "Will you work for us in Moscow?" he asked. The CIA would pay him $25,000 a year, the money to be deposited in a Western bank account.

Nosenko weighed the offer in his mind. He had one condition, he declared after some thought. There must be no attempt to contact him in Russia. It would be "too dangerous."

Both CIA officers concurred; that had been their assessment, too. And without any further discussion, it was settled. Nosenko agreed to become a paid American agent.

As soon as the Russian left that evening, Pete sent a succinct cable to headquarters: "Subject has conclusively proved his bona fides. He has provided info of importance and sensitivity. Willing to meet abroad."

It was understated, but its tacit message would be read loud and clear in Langley. Pete had closed the deal.

Their final session two days later had the sentimental feel of the last day of term. But instead of a diploma, Nosenko was presented with a secret writing kit in case he needed to send an urgent letter from Moscow. Pete advised, however, that it would be prudent to wait until KGB business took him back to the relative safety of the West before communicating. Then he should send a telegram signed "George" to the address Pete proceeded to share. Two days later he would be met at precisely 7:45 p.m. in front of the first movie theater listed alphabetically in the phone book of the town from where he'd sent the cable.

With the housekeeping out of the way, there was a round of toasts, and still another as everyone got deeper into the valedictory mood, and finally one more because there was no longer any reason for restraint. But then it was time. First Pete, then Kisevalter wrapped their arms in hearty Slavic bear hugs around the Russian, and after a wave goodbye and a heartfelt wish of good luck, Nosenko trundled off. The next day he boarded a plane to Moscow to begin his new life as a CIA mole in the KGB.

Later that same day, before Pete had barely a moment to relish his triumph, a message arrived from Langley. He and Kisevalter were wanted at headquarters. *At once.* And there was one further instruction: They were to take separate flights, each carrying a duplicate set of the interview tapes. Headquarters wanted to make sure that if the vengeful Russians waylaid one of the officers, there'd still be a good chance that the other would make it across the Atlantic to deliver the prized goods.

"Is he for real?" challenged Jack Maury, the head of the Soviet Division, only moments after the two jet-lagged officers were seated in his fifth-floor office.

Pete decided to let Kisevalter answer; he was the senior officer. "There's no sign to the contrary," Kisevalter replied carefully. "He sure talks the way only a KGB man could."

But Maury kept pressing. "But why the hell did he take that kind of risk for a few hundred bucks?"

"I don't know," Kisevalter conceded. "Pete and I have gone all around the barn talking about it, and we still haven't come up with an answer."

Maury sat imperially behind his desk, and the two officers were seated opposite him on an enormous sofa. It was over eight feet long and had been custom built years ago for a defector who'd insisted he did his best thinking lying on his back, his long legs stretched out. The running joke in the SB was that this piece of bespoke furniture was the only contribution the recruit had ever made to Western intelligence.

Up to now Pete had listened quietly, but all the time he'd been growing increasingly apprehensive. He didn't like the way the conversation was going. Hadn't Maury read the transcripts? Pete saw they lay at his elbow. Was the fifth floor really going to look the proverbial gift horse in the mouth? Don't they realize the opportunity? *A mole, our mole, prowling the corridors of Moscow Center.*

"There must be more to this than a few bucks," Pete agreed evenly, finally speaking up. If the conversation became confrontational, he feared the fifth floor would do what they always do when things got heated: accuse the naive fieldman of not seeing the Big

Picture. Hoping to move things forward, he offered, "Well, we'll take a good shot at that next time. For now we'll just have to live with it." He had meant to be conciliatory, to assuage Maury's concern about the flimsy motive that had prompted Nosenko to come running for help, but the words came out with more edge than Pete had intended. He worried that he'd overplayed his hand.

"And count ourselves lucky," Maury, nodding his head in agreement, finally replied. And with that, Pete knew the battle was over. He'd won.

MAURY'S SECRETARY ENTERED, CARRYING A tray loaded with a well-polished silver-colored coffeepot and three china cups and saucers. Playing the gracious host, Maury poured, and Pete and Kisevalter shared a conspiratorial wink. So this is how things are done on the fifth floor, it signaled. No cafeteria Styrofoam for our high-flying leaders.

The coffee break, though, was short-lived. Pete had barely taken a sip from the delicate cup before Maury abruptly put things back on an operational footing.

He had a plan. As he shared it, it grew clear to Pete that his boss had never truly doubted Nosenko, or his potential. Maury had merely been toying with them, probing to see if, when pressured, they'd provide the reassurance he required.

"Career management," Maury called it, and the intentional irony made him smile. The gist of the scheme, as he laid it out, was simple: The CIA would advance Nosenko's career in the KGB. They'd provide him with the sort of twenty-four-karat intel about CIA ops that would make his Moscow Center bosses think Nosenko was the greatest secret agent since Feliks Dzerzhinsky. It'd be costly. The scheme wouldn't work unless productive agency ops were blown to smithereens. But in the long run, it'd be a price worth paying. The higher Nosenko climbed up the ladder in the KGB, the greater the potential reward.

Then Maury suddenly seemed to be in a hurry. There was another meeting he needed to get to; he was already late, he complained. He was nearly at the door, and Pete was following in his wake, when he abruptly turned and spoke. It was as if the idea had just occurred to him.

"Before you leave, Pete, you'll want to look into some new information we've got. There's been an important defection from the KGB. He's here in Washington," Maury explained.

Pete nodded. He was immediately interested.

"And do check in with Jim Angleton," Maury added. "He's aware of the Nosenko contact with us, but he'll want to have your details."

I bet he does, Pete thought, but decided not to say anything.

They were in the corridor, Maury rushing off in a self-important whirlwind to wherever it was that his presence was immediately required, when once again he turned, struck by another new thought. It concerned the recent high-level KGB defector, the one who was lying low in Washington. There no longer seemed to be any reason, he said, to have the overworked SB analysts go hunting for the file. "You might as well get it from Jim," he instructed breezily, before disappearing down the long hallway.

And all Pete could do was wonder what that was all about.

Chapter 10

Like many people at the agency, Pete had *his* Angleton story. Only unlike so many of the other tales, Pete's wasn't apocryphal. He'd witnessed it firsthand. In fact, he'd been smack in the middle of it. And while he made no claim that, in the scheme of things, it had been an incident of great consequence, the always observant fieldman had nevertheless filed it away. It revealed, he believed, a bit about how Angleton's mind worked. The problem, though, was that Pete couldn't fathom precisely what it conveyed. The whole thing had left him puzzled. Was Angleton just being a tease? Was it a test? A joke? Or was it proof that the world going on in Jim Angleton's head was an entirely different place than anywhere mere mortals lived?

It had been during Pete's last posting back at headquarters, before the shiny new campus had been spread across suburban Langley and he'd been a young deskman supervising operations against Polish Intelligence. Not long after he'd arrived back in Washington, he'd been brought into Angleton's circle by William Hood, who'd served in X-2 with Angleton during the war and had been Pete's station chief in Vienna. Hood's imprimatur was one thing, and the fact that Pete, smart, well connected, and daring to boot, was an officer clearly going places at the agency was another. Soon Pete, the youngest by at least a generation, was to his delight being asked along for lunches where Angleton held court, OSS warhorses traded tales of past glories, and rivers of martinis flowed well into the afternoon. He was also invited to jovial, well-lubricated dinner parties at Angleton's rambling old colonial on Thirty-Third Street in North Arlington, a relative youngster among the guests, who tended to be well-bred Ivy League spies who had done interesting things during the war.

The usual practice at these Saturday night gatherings was, after the cheese and port had been served, to adjourn to the living room for a spiritedly competitive game of charades. And so it happened that one evening Pete was offered a challenge by the host. On a slip of paper, Angleton facilely wrote out an obscure line from, Pete would later learn, the second movement of T. S. Eliot's *Four Quartets*: "clot the bedded axle tree." Next, all a befuddled Pete had to do was successfully convey the abstruse phrase—visionary? epigrammatic? gibberish?—to the others.

Pete, always up for a challenge, had struggled valiantly; and yet in the end he'd only succeeded, he felt, in making himself look silly, even uneducated and unworldly, in front of an audience of senior agency officers. In the aftermath, he'd sulked for days. What had Angleton been trying to do? Humiliate him? Or did the phrase carry some sort of secret meaning that Angleton, in his cryptic way, had been trying to convey? Pete never arrived at a satisfactory answer.

While not entirely forgotten, the incident was not on Pete's mind when, not long after leaving Maury's office, he followed up on the order that he brief Angleton on the meetings with Nosenko. The head of Counterintelligence listened without interruption while all the time busily drawing an elaborate geometric design on his pad with a pencil. So Pete plowed on to the end, concluding with an upbeat assessment for future product from Nosenko. Then he waited for Angleton to chime in.

Angleton's unnerving silence stretched on. Finally he dropped his pencil into his out-tray, studied his doodle with some concentration, and tore it into pieces. A moment later he methodically deposited each of the scraps into the classified burn box adjacent to his desk. When this ritual was completed, he broke his silence.

Without further explanation, he reiterated Maury's advice: Pete should read the files on the new KGB defector. His name, Angleton revealed, pronouncing each syllable with deliberate precision, was Anatoly Mikhailovich Golitsyn.

Angleton opened a safe across from his desk and extracted a tall stack of buff-colored folders. But before he handed them over, he summoned his secretary into the room. Take Pete, he instructed Bertha, to the "counterintelligence conference room." He'll need to concentrate, Angleton added, the words a firm command.

The conference room was across the hall, and Pete quickly decided it'd be more suitable for use as a closet. It was tiny, windowless, and the walls had been painted a weary, slush-colored beige. The only furnishings were a government-issued metal table and a couple of straight-backed chairs. But once Pete started reading, he was transfixed.

As the hours passed, he found himself without warning suddenly traveling back in his mind to that challenging game of charades he'd played years earlier. In his primed state, he saw, or so he wanted to believe, the implicit path that lay between the arcane line of poetry Angleton had chosen and his present laborious pursuit. It had not been a random bit of Eliot. Rather, Pete now understood that Angleton, in his indirect way, had been sharing a counterintelligence maxim: the need to draw ineluctable connections, to fit seemingly disparate pieces together. For the extensive Golitsyn files, he grew to realize, were the "bedded axle tree," the firm connecting rod, that would allow him to make sense of what Nosenko had been up to when he'd first come knocking on the safe house door two weeks ago in Geneva.

THE GOLITSYN FILES TOLD, PETE began to perceive as he continued to make his ruminative way through the hundreds of pages, two distinct stories. The first was the more accessible tale. And here even the circumspect prose of the agency's daily activity reports could not deaden the adventuresome narrative.

A doorbell rings on a snowy evening in suburban Westend, a wealthy enclave of imposing homes just outside the Finnish capital of Helsinki. Frank Friberg abruptly stops shaving; it's ten days before Christmas 1961, and he's getting ready to head off to a holiday party. Who could it be? he wonders. He wasn't expecting anyone. Besides, only certain people know where he lives; his address is secret. At that moment, his thoughts are orderly, an attempt to rein in a rising sense of danger. Another loud and insistent peal from the doorbell blares like a trumpet through the evening quiet. The CIA station chief at last puts down his razor; removes his revolver from the nightstand, concealing it in the pocket of his robe; and then, with the unnatural calm that pros affect when there's no turning back, answers the door.

Three faces stared back at him: a man built like a fire hydrant, short and thick; a taller woman with flaming red hair; and a young girl, her small hands cradling a doll. The falling snow had left them covered in white, a spectral trio. Friberg's first thought was that they were Christmas carolers, and he removed his hand from the grip of the gun in his pocket. Then he noticed the fire in the man's eyes: It was the wild glow of fear.

"Do you know who I am?" the man asked. His voice was a beseeching whisper.

Friberg had no idea. He had opened his door only a few inches, continuing to use it as a shield.

"I am Anatoly Klimov," the man announced.

At once Friberg motioned for them to come inside, and then swiftly shut the door behind them. Friberg knew the name well: The agency watchers had identified Klimov as a KGB officer working under diplomatic cover at the Soviet embassy in Helsinki. When the Russian removed his fur hat, Friberg had a better look at the man and matched the jowly face with the photograph in his office files.

They were sitting in the living room, and while Friberg spoke no Russian and Klimov only a smattering of English, the KGB man conveyed that the redheaded woman was his wife, and the girl holding the doll was his daughter.

Then the Russian removed a scrap of paper from his suit pocket and began to write. With some difficulty, he managed to spell a single word: *asylum*.

Friberg refused.

He had been with the agency since he'd graduated from Harvard a decade ago. There had been a long tour based in Sweden where under commercial cover he traveled across Europe, an itinerant tough guy thrown into sticky situations. But that kind of work takes its toll on an agent, and headquarters judged he could use a rest. A diplomatic cover job was found at the embassy in Helsinki. Friberg spoke the language, kept his head down, and after a few years he was promoted to station chief. It was easy duty, far from the Cold War battlefields of his earlier service. But Friberg was a pro of the old school, and he knew how things worked: you don't give a KGB hood asylum; you keep him in place, run him as a double. He started into his pitch.

Klimov cut him off. He refused to discuss the possibility. Instead, he explained—"lectured," was actually more like it, Friberg would later say with a hint of admiration—that he'd been planning this for the past year. He'd been waiting until his daughter returned for the holiday break from her boarding school in Moscow; they'd leave as a family. But, he sternly insisted, there was only a two-hour window to get them all out of Finland. A minute longer and he was certain the KGB would notice his absence. Once they'd sounded the alarm, he'd never get out alive.

Friberg had to make a decision, and the clock was ticking. For a moment he considered the possibility that his guest was exaggerating his predicament. But the iron rule was that the Joe always calls the shots; he can best appreciate the dangers he's facing. And with that maxim firm in his mind, years of tradecraft took over and he jumped into action. He called the airport and learned that in two hours there was an eight o'clock flight to Stockholm. He made up his mind to get the family on board.

There was no time to manufacture fake passports; he'd need to improvise. The best he could do, he decided, was to get US visas affixed to their Russian passports. Another call and this time to a CIA colleague and the officer agreed to rush to the airport with the embassy stamping equipment. The plan was to meet in the men's room on the main concourse, and he'd stamp the three passports in the protective privacy of a toilet stall. The tickets were booked in the name of Klements. Friberg hoped the name was close enough to Klimov so that a Finnish passport control officer wouldn't notice the discrepancy, while at the same time it was enough of a twist and wouldn't jump off the page if the Russians were scanning flight manifests.

Friberg's car was parked in the driveway, and he made sure to turn off all the lights, both inside the house and lining the front path, before leading the trio to his Volvo. If any Russians were lurking about, he wasn't about to give them a clear shot.

As they hurried in the dark to the car, Klimov suddenly broke away. For an unsteady moment, Friberg thought that maybe the Russian had a change of heart, or perhaps the entire evening was an escapade staged to embarrass the CIA and smirking photographers would jump out of the bushes. But Klimov was only digging up a package out of the snow.

Once he was in the car, the Russian explained he had buried it before ringing the doorbell. It contained documents he'd smuggled out of the Soviet embassy, he went on, clutching the package to his chest. He'd hand it over when he and his family were safely in America.

The drive to the airport was another trial. A race against the clock over roads blanketed with fresh falling snow, and the looming threat of Russian thugs waiting to pounce around every bend. As Friberg drove at a dangerous speed, his eyes darting from the slippery road to the rearview mirror, the Russian shared two confessions. For one thing, he was a counterintelligence officer; his primary operational target was the United States. And for another, Klimov was a workname. His actual name was Anatoly Mikhailovich Golitsyn.

The group managed to get aboard the eight p.m. flight to Stockholm just before it took off. They were three abreast in coach, and Friberg sat across the aisle, constantly on alert. Four tense days later, after a sputtering trip designed to evade any pursuers sent by Moscow Center, they arrived in Washington. The next day the family was settled into a five-star agency safe house, a well-appointed columned mansion in the Virginia hunt country.

In the busy weeks that followed, an honor roll of CIA officials showed up at the hunt country manor to pay their respects and to listen to what Golitsyn had to say. The attentive audiences included a deputy director of the agency, the head of the Russia Division, and, most significantly, as things proceeded to play out, the counterintelligence chief, James Angleton. The transcripts filled volumes.

Now two years later, as Pete remained glued to his uncomfortable metal chair in the bleak conference room, he moved into this second part of the case history, the interrogations. What he read in the debriefs shook his world.

I'VE HEARD THIS BEFORE, PETE began to realize with growing discomfort as he continued to make his way through the Golitsyn transcripts. The defector recounted the *identical* incidents and operations that Nosenko had shared in Geneva. The coincidence was unsettling. Sure, Pete tried to persuade himself, it was possible that two KGB officers from different departments would know a bit about what their colleagues were up to. After all, that's what spies do: They poke into things they shouldn't. But even as he tried out that argument, it immediately collapsed. Specific details of clandestine operations wouldn't be bandied around the samovar, especially in a fortress like Moscow Center where the rules were stern, inviolable decrees.

But what really had Pete's suspicions swirling was that these two overlapping accounts, he couldn't help but notice, arrived "almost on the heels of one another." And in Pete's profession there was no room for coincidences.

Methodically, he began to compare the two versions.

He learned that there was nothing earth-shattering in Nosenko's sharing the ingenious surveillance techniques of the Second Directorate watchers. It was old news. Golitsyn had spilled those beans two years earlier. Same for the identities of many of the agents in place Nosenko had revealed Moscow Center was running. Golitsyn had already stripped them of their covers.

And Golitsyn, too, had focused on the KGB's detection of Popov. Only he'd told a different story. According to Golitsyn's telling, Popov had been blown much earlier than in Nosenko's account. And there was no mention of the KGB watchers playing a part in Popov's downfall.

Yet why had Nosenko been so set on telling his story? Pete wondered. Why give a different date for when Popov had been tumbled? Why the story about the KGB spotting a US embassy official mailing a letter to the double agent? It made no sense.

Unless he'd been trying to send us in the wrong direction. Unless he'd been trying to hide something.

But what?

The more Pete read, the more his clutter of doubts, uncertainties, and suspicions hardened into a firm thesis. Nosenko's stories were a palimpsest of Golitsyn's. Only Nosenko had written over his predecessor's accounts with an artful maliciousness. His deliberate intention, Pete had come to see, was "to dismiss or divert suspicions that the earlier reports had evoked."

He now knew: Nosenko was bogus. A dispatched agent.

But why? Pete again demanded. What were the Russians up to?

It was a dangerous question because he knew the answer would be even more dangerous. It would mean the KGB had shrewdly sent someone from their own ranks to spread disinformation that would divert attention from a very powerful secret. But what secret? What was worth hiding that Moscow Center would go to the trouble of sending Nosenko to impugn Golitsyn?

Spurred on, continuing to dig deeper into the volumes of files, Pete found the answer he'd been fearing. It was in one of Friberg's debriefs. The agency interrogators were coaxing him to recall all that Golitsyn had said during the meandering journey from Stockholm to Virginia. "He said," Friberg suddenly remembered, "he had seen some material at KGB headquarters that could only have come from a very sensitive area of the agency. From a place high inside the CIA."

PETE SLEPT ONLY INTERMITTENTLY THAT night, his restless mind attacked by disappointment and newfound fears. In the morning, he went straight to Angleton's office.

"Thanks, Jim," he began. "You were right. I needed this information. But at the same time, I've got to tell you something. We may have a problem."

Pete shared his burgeoning theory about, as he put it, "the curious coincidences and the persistent overlapping" in Nosenko's and Golitsyn's accounts. He was careful, though, not to take things too far.

Angleton frowned, and then shook his head. But whether he was saying no to Pete's view of things or simply expressing his dismay at the machinations that take place in the covert world was not revealed. All Angleton said was, "Please jot down these points for me. I want to look carefully at this."

The next day Pete handed Bertha an envelope that contained

a handwritten list. Pete had jotted down fourteen points of parallel reporting in the two Russians' accounts. He could have added more, but he'd decided this would be sufficient. It was a tacit indictment of Nosenko, and once again he didn't want to overplay his hand.

Somehow Pete managed to pass the hours until he was summoned. What he did, whom he spoke to—Pete had no idea. The details would always remain a blur. Pete's entire being had been focused on the prospect of Angleton's response.

"You may be on to something here," Angleton said guardedly just moments after Pete entered the dark office.

Pete let out a silent cheer.

"As a matter of fact," Angleton continued, "I agree with you. Golitsyn himself said he expected the KGB to make some effort to divert the leads he could give us. Maybe that's what we got on our hands now."

Only Pete suspected there was something more. *Golitsyn had seen some material at KGB headquarters that could only have come from a sensitive area of the agency. From a place high inside the CIA.* If he was right, they had something a lot more dangerous on their hands.

Chapter 11

1964

MEANWHILE, BACK AT DERIABIN'S MATCHMAKING dinner party that was the center of Pete's present world, the shared recollections of the Nosenko case had gathered steam. And the conversation turned to Nosenko's reappearance. Only this time he'd manage to involve not just Pete but all the old spies seated at the table in the ensuing drama. But perhaps that was to have been expected, given the circumstances: Nosenko had put himself smack in the middle of the biggest mystery of the twentieth century.

THE STORY OF NOSENKO'S RETURN, as well as Pete's involvement in this new stage of the case, was very much a story of the times. And to understand that period at the CIA, the rocky, anxious mood that stirred the agency as well as pretty much the entire country, the place to start was on November 22, 1963.

People would always remember where they were when they heard the news that President John F. Kennedy had been shot, and for Pete that shock came as he was entering an elevator at Langley. For nearly two months he'd been based at headquarters, running counterintelligence for the Soviet Division, and from the start he'd been knee-deep in trying to get a handle on the ominous questions arising from one blown op after another. But that afternoon the axis of his world abruptly shifted.

Returning from lunch, Pete had one foot in the elevator when Jerry from the fifth floor stepped out and, his voice a shrill keen, remarked, "Isn't it terrible." Pete's lunch companion bantered back with snark. "Probably not as bad as all that, Jerry."

"Haven't you heard?" Jerry implored. "The president has been shot in Dallas."

The days that followed were, despite the mournful mood, a whirlwind at the agency, and the Counterintelligence Staff was at the operational center. It served as the focal point for the cascade of facts, theories, and half-truths pouring into the clandestine services challenging whether a self-proclaimed Marxist named Lee Harvey Oswald, using a $21 mail-order rifle, was solely responsible for the assassination. And Rocca, Deriabin, and Pete were quickly thrown into the eye of the raging investigative storm.

Within a week of Kennedy's death, President Johnson established the Warren Commission to get to the bottom of things and, with Angleton deftly pulling the strings in the background, the Rock became the agency's liaison to the probe. Yet it was a very circumscribed collaboration. For all the while that he was running point between Langley and the commission gumshoes, Rocca was guarding as best he could a dangerous secret.

The CIA had dutifully disclosed a surveillance report in its files that documented Oswald's showing up at the Soviet embassy in Mexico City just weeks before the murder in Dallas; he'd applied for a visa and met with "consular officials." However, what the agency—and Rocca specifically—were not sharing was an even more provocative detail in that already provocative trace: One of the diplomats who'd huddled with Oswald was Valery Kostikov. When his name had been fed into the CI computer, red lights had started flashing. He was identified as a particularly disagreeable member of the Thirteenth Department, the KGB unit that busied itself with "wet works," as murders were delicately referred to in the trade.

Was this the incriminating piece of intelligence that would bring the artfully constructed theory about a misfit lone gunman crashing to pieces? Did it tie the Soviet Union into the plot? Or was it just another odd coincidence in a quixotic world? Rocca and his bosses decided that they, and they alone, could best answer those questions. They believed that it was in the national interest to hold on to inconvenient facts. "A benign cover-up," the CIA's in-house historian would meekly chastise years later.

Yet while the Rock was busily firming up the agency's party line that Oswald had acted on his own, Deriabin, the defector now a trusted analyst on the CIA's payroll, looked at the events in Dallas with his own well-versed gaze. And he grew convinced he understood precisely what had happened. Five days after the assassination, he sent off a stridently confident eight-page typed memo that landed at the agency with the disquieting rumble of a ticking bomb. Hammering away, he proclaimed with one vehement assertion after another that his old KGB bosses had "instigated our president's death." He insisted that Oswald was precisely the sort of "manageable type" Moscow Center would've recruited. He even found support for his theory in the fact that Oswald had been arrested in a Dallas movie house; "certainly we know the KGB's penchant for using theaters for meeting places," and therefore it was likely that Oswald had shown up to meet his handler. (At the same time, it might have occurred to the agency's anxious Soviet Division officers reading Deriabin's letter that this was their tradecraft, too; Nosenko, for one example, had been instructed to appear at a movie house if he ever came back to operational life.)

Pete was also busy. Just hours after the president's death, his CI team unearthed a photo taken by an American tourist that, thanks to the pack-rat mentality that is at the core of intelligence work, had been buried in an innocuous file headed "Minsk Landmarks." In one of the amateur's shots there he was: Oswald, the now unforgettable smirk planted on his face, caught by happenstance in front of the Minsk opera house. In the early hours of the investigation, it was, improbably, the first tangible proof that the suspect had spent time in Russia. And just a day after the assassination, he personally sent the commission a transcript of Oswald's October conversation with the Soviet embassy in Mexico City (recorded surreptitiously by a CIA wire team) to supplement the surveillance report found earlier in the SB's files. Further, as the inquiry gathered speed, Pete, in another memo sent to the Warren Commission, let it be known that both Rocca and he were of a single mind: they thought there'd be a lot to be gained if Deriabin was present when the FBI interrogated Oswald's widow, Marina. They believed that the canny old KGB man would be able to get a handle on whether the marriage was the love story her supporters claimed, or whether it was the pragmatic linchpin of a Moscow Center–controlled intelligence operation.

Then, as the analytical machinery at the spy agency sputtered on, Pete, without warning, found himself thrown into the heart of the mystery. On the wintery morning of January 23, 1964, he was summoned by David Murphy, the new, ambitious head of the Soviet Russia Division, a veteran who had arrived at Langley fresh from a grueling tour as the Berlin station chief. "Come in, Pete," Murphy ordered over an internal phone line. "There's news," he added, the words striking Pete as cryptic yet buoyant.

The division heads had corner offices on the fifth floor, and Pete walked down a long corridor to reach Murphy's hard-won real estate. He entered and saw that George Kisevalter, a self-satisfied smile on his broad face, had already taken a seat on the couch.

"Nosenko's back," Murphy quickly announced, as though rushing to beat Kisevalter to the punch.

"The telegram came in last night," Kisevalter explained, positively beaming. "It's Geneva again."

Excited, Pete threw a boxer's short right jab against an imaginary jaw. It was an instinctive gesture, but he was also looking forward to a fight. The Golitsyn files had radically transformed his initial appraisal of Nosenko. This go-around with the Russian would be a more combative session.

Murphy laid down the operational ground rules. Secrecy must be paramount, he lectured. There'd been too many leaks recently. The two officers must not tell anyone where they were going, or why. Book separate flights, he continued. We don't want anyone to suspect you're on a joint mission. He concluded with the instruction to cable him directly; the fewer eyes on this, the better.

Pete hurried home to pack. Geneva, he knew, would be cold and wet in January.

The ABC Cinema was at the top of the list of the movie houses in the Geneva phone book, and if Nosenko stuck to the operational rules that'd been established eighteen months earlier, at 7:45 he'd be standing outside.

A half hour before the appointed time, Pete got off the bus that had taken him into the busy commercial center of the city. He'd deliberately exited blocks away from the cinema. The night was frosty, but he walked a long, haphazard route, pausing at shop windows as his eyes furtively checked out the passersby mirrored in the glass.

Pete had spent many years in Geneva, first as a student and later as a spy, and there was a chance that someone would recognize him. In a hasty attempt at a disguise, he wore the horn-rimmed glasses he normally used for reading, and he had purchased a black Styrian hat, tugging the wide brim rakishly low on his forehead.

He made it to the meet at 7:45, and for once everything seemed to be proceeding as planned. There was Nosenko standing to the left of the ticket booth, peering at his watch impatiently, another moviegoer growing cranky as he waited for his tardy date.

Pete's next move would be difficult. He couldn't walk up to

Nosenko, pat him jovially on the back, and then lead him off to the new safe house the local team had arranged. Pete was confident he was clean, that he'd arrived without any enemy watchers on his tail. However, there was no telling whether the KGB had put a surveillance team on Nosenko, and the rule was to always expect the worst.

A brush pass is a ploy that's all about speed and naturalness, the fieldman's art of handing off a canister of film or a purloined document in a crowd without attracting attention. Back in the game, Pete's old skills kicked in. He scurried past Nosenko, thrust a paper with an address and a phone number into his hand without breaking stride, and then continued on down the dark street.

By the time Pete, having taken care to shake any watchers, made it to the posh new safe house in an upscale residential neighborhood of the city, Nosenko was already in the living room. He was standing across from Kisevalter. Pete had the feeling he'd interrupted a heated conversation.

"Yuri has a bit of a surprise for us," George explained uneasily. "He wants to stay."

"What?" Pete burst out. He tried to persuade himself he was not understanding things. "You don't mean defect, do you?"

"Yes," said Nosenko firmly. "And right now. I don't want to go back."

Pete was flustered. The last thing he wanted was to bring home a defector he didn't trust. His plan had been to beat the bogus agent at his own covert game. He'd run the Russian in place, yet all the time he'd be feeding Nosenko disinformation that'd have his KGB masters chasing down blind alleys for years to come.

"We'd better sit down and talk about this," Pete tried. "And let's have a drink—I know I could use one," he added, at last speaking with complete honesty.

A plate of smoked salmon surrounded by wedges of a hearty black bread was on a tray, and as Nosenko helped himself, Pete poured the scotch. While they ate and drank, Pete's mind churned. He needed to regain control of the situation.

"I don't understand," he said at last. "You said you would never leave your country and your family. Is something wrong?"

"Yes. I don't know exactly what, but I've been getting the feeling that they might be on to me."

This made no sense to Pete. If the KGB had any suspicions, they'd never have sent Nosenko back to Geneva as a security officer for the new round of the disarmament talks. And it made even less sense if Nosenko was the dispatched agent Pete was certain he was.

"But I still don't understand," Pete suggested mildly. "What about your family, the little girls?"

Nosenko was unruffled. "Oh, they'll be okay."

Pete knew, though, what would be in store for Nosenko's family. They'd be kicked out of their Moscow apartment and resettled in some bleak village in the far reaches of Russia, forever tainted by their relationship with a traitor.

If he was, in fact, a traitor.

Pete needed to buy time. He had to work out a plan that would squeeze some operational value out of Nosenko—and then pull the plug on the madcap idea that the Russian would be welcomed as a defector.

"Stay where you are, at least for a few days," he suggested helpfully. "If you sense any real danger, you can come here any time. They won't kidnap you in Switzerland." But Pete was thinking: kidnap? The KGB doesn't snatch its own active agents.

"Maybe for a week or so," Nosenko said, pursing his lips, as if to emphasize the turmoil this concession was causing.

Finally, to Pete's great relief, they settled into the debrief. Only no sooner had Nosenko begun talking than he completely demolished Pete's carefully plotted plan.

Not bothering with any preliminaries, Nosenko dove right in. He announced that he'd been personally involved in the KGB's investigations of Lee Harvey Oswald, both before *and* after the Kennedy assassination.

"When Oswald came to the Soviet Union in 1959," Nosenko recounted with a matter-of-fact fluidity, "he told his Intourist guide he wanted to stay in the country and become a Soviet citizen. I was deputy chief of our section dealing with American tourists in the USSR. And I judged that Oswald was of no interest and would probably just be a nuisance. So I decided to reject his request."

Pete wanted to interrupt to say that it was extremely unlikely that a midlevel KGB officer like Nosenko would be the one making such a significant decision, but he decided that for now it'd be better to let the tale play out.

"When Oswald was told he couldn't stay," the Russian continued in the same steady voice, "he went back to the hotel and tried to commit suicide. They found him in his room with his wrist cut. Higher-ups decided it would be too embarrassing if he should really succeed in killing himself in our country. But he would not be allowed to stay in the Moscow area. The Red Cross found him a job in Minsk."

At this point, Pete attempted to add some soda water to Nosenko's drink; he needed to keep the Russian focused. But he was brusquely waved off. Defiant, Nosenko poured a few more inches of scotch into the glass.

"Didn't you even suspect he might be an American spy?" Pete asked, deciding he'd better speak up before the would-be defector finished off a bottle of single malt and drifted off. "Didn't any KGB people at least interview him to get their own impression? To see if he might be useful to the KGB? After all, he had just left the Marines."

"No. No one even bothered," Nosenko repeated. "He was obviously just a low-level corporal or something." Pete noticed that the Russian's gaze remained as direct and innocent as before.

Abruptly, Nosenko's account now picked up speed. "Then came the news about his assassination of President Kennedy. The

Americans might get the idea that we had something to do with it. So Khrushchev himself asked my boss if the KGB had anything to do with Oswald. Immediately my boss told me to get the KGB file from Minsk. I phoned Minsk, and they flew a man right away with their file on Oswald.

"I personally reviewed it to see if the KGB there had been any relationship whatever with Oswald," he said proudly.

"*And?*" Pete challenged.

"And nothing. There was no sign whatsoever that the KGB in Minsk had taken any interest in him."

"Didn't they watch him, or bug his apartment?" Kisevalter coaxed.

"No, nothing of the sort."

Pete tried another tack. "What did the file look like? How big was it?"

Nosenko spread his thumb and index figure about an inch apart and said, "One volume. Thick like this."

"Did you read it all?"

"Had to," he said mildly. "I read it carefully."

Nosenko fell silent, a deliberate, cagey pause. Then he resumed: "If anyone wants to know whether the Soviet government was behind Oswald, I can answer—it wasn't. No one in the KGB paid any attention."

Pete appeared to be not quite listening, but his mind was shouting, *If anyone wants to know?* Just the Warren Commission and the CIA. Not to mention a couple of hundred million Americans.

And all the while Nosenko kept talking. He had still another revelation up his sleeve. One more fortuitous coincidence.

"A few months ago, long before the assassination," he went on, "I happened to be visiting an office in the First Chief Directorate. A cable had come in from the residency in Mexico City. They showed it to me. This guy Oswald had come into the consulate saying he'd lived in the USSR and wanted to go back. He was asking for a visa to return here."

"How long was the cable? Exactly what did it say?" Kisevalter interjected.

"About a half a page, no more." Then Nosenko swiftly went on with his account. "I heard the guys talking it over. They decided there'd be no good reason to let him come back. So they sent a cable telling Moscow to refuse the request."

The fieldman in Pete was now on full alert. He needed to get some perspective on what he was hearing. Perhaps, he suggested, Nosenko could use a break. But, in truth, it was Pete who had to get himself grounded.

The tray of salmon was replenished, another round of drinks was poured, but these were just distractions, and feeble ones at that, to the raging internal chaos that Pete was trying to calm. He had been running the agency's only source inside the entire KGB, and now this officer had returned after a long absence to announce that he had not one but four separate points of contact with the Oswald affair. No less extraordinary, he had personally reviewed the entire Oswald case file and could testify that Moscow Center had absolutely no involvement in the murder of the president.

Like everything about Nosenko, it was all too good to be true. It strained credulity. All Pete's professional instincts were signaling that Nosenko had recited a carefully written and well-rehearsed script. At the same time, Pete understood that there were two equally plausible explanations for the Russians having dispatched Nosenko with this dubious information. Either it reinforced the truth: The KGB had no role in Kennedy's assassination. Or, more chillingly, Nosenko had been primed to deliver a pack of lies designed to cover up Moscow's hands-on participation in the shooting of an American president.

The only certainty, Pete had dolefully come to realize in the course of this astonishing evening, was that Nosenko would get his wish: he'd be allowed to defect. Pete's concerns about the Russian's truthfulness, of his being a dispatched agent, had become irrelevant. Nosenko couldn't be allowed to go back to Moscow. He would have to share this important story with others at the agency, perhaps even with members of the Warren Commission. Its truth or falsity needed to be definitively determined. By the rules of the trade, Pete had landed a glittering prize. It didn't matter that Pete personally put little stock in anything he'd heard. For the moment, the only question that remained was how long it would take to activate the defection protocols.

As for the rest of the debrief that long day, Pete, his head a jumble of conflicting duties, had little recollection of what was discussed. Fortunately, the recorders were running and they captured what transpired, even down to the splashes in Nosenko's glass as he poured a flood of scotch. As the tape winds down, there's the distinct sucking sound of Nosenko draining his glass, and his uncharacteristically weary voice announces, "If I'm going to stay with the delegation for these next days, I'd better go."

FOUR DAYS LATER, NOSENKO APPEARED without warning at the safe house. He was agitated, a man acting as if he was in considerable danger. A telegram, he claimed, had arrived that morning at the Soviet mission. He was being recalled to Moscow.

"This must mean they've found out about our contact," he wailed to Pete. "Look, I have to leave, right now. I don't even want to go back to the hotel."

Either Nosenko was lying or the threat was genuine, and Pete knew he couldn't take the chance on which was the reality. The legal arrangements had not yet been finalized to bring Nosenko to America, so Pete improvised. He swiftly commandeered an American officer's uniform and Nosenko, enjoying the masquerade, all his previous agitation gone, put it on. Within hours, Nosenko received the sharp salutes of the border guards as his chauffeured army sedan crossed into Germany.

Dave Murphy, fresh off a plane from Washington, met with the Russian the next day at the CIA station in Frankfurt. The head of the agency's Soviet Division wanted to take his own measure of the KGB man who had such a remarkable story to tell. He listened to the details with attention, but what struck him most was Nosenko's composure during their interview. It was as if, he'd later say, Nosenko had been expecting to meet him.

On February 11, 1964, Nosenko, accompanied by a burly CIA watchdog, flew to the United States. His plane landed in New York City on a gray, bone-chilling morning. The airport had been renamed six weeks earlier in honor of John F. Kennedy, the assassinated president.

Chapter 12

THE BAR WAS A SHORT stroll from the lights of Baltimore Harbor, discreetly tucked away from the downtown hubbub and set back near the end of a long, lonely street. Inside, the room was dark, the patrons ran to serious drinkers, and the music from the jukebox was a steady blast to which no one seemed to be paying much attention. The agency security team had chosen this forlorn watering hole because they'd hoped it would be too down-market to attract any of the Russian embassy crowd from DC, even the more adventurous ones who might set off to explore new, raunchy haunts.

But perched on a stool near the very end of the wooden bar top, a scotch gripped in his hand as if for balance, Nosenko was in his element. He'd spent a long evening bellowing at the bartender, ostentatiously ogling any female in sight, and methodically drinking himself into a stupor. That is, it was a night like all the others.

So perhaps the two burly agency babysitters' lack of attention can be excused. They were huddled in a back booth, nursing coffees and griping about the boorish Russian drunk it was their unfortunate duty to chaperon, when the fight broke out. As the rapidly unfolding events were later reconstructed, Nosenko had reached out and firmly pinched a shapely passing barmaid on the butt. She let out an annoyed shriek, and a longtime admirer jumped up from his stool to defend her honor. A boozy windmill of a punch was thrown that barely managed to graze Nosenko's shoulder. The Russian, despite his inebriated fog, retaliated with an impressively swift and accurate combination of jabs; he'd been an amateur boxer in his youth. By the time the security officers had made it to the front of the bar, the chivalrous patron was stretched out cold on the floor, the barmaid, her hands clenched into tiny fists, was pummeling an amused Nosenko, and the bartender was shouting that the police were on the way.

Things quickly teetered toward disaster. The cops had their handcuffs out; the two agency men had their hands menacingly close to the weapons under their suit jackets; and only the arrival of a Baltimore police captain, who'd been awakened from a sound sleep in his own bed by an imploring call from a CIA night-duty officer, had convinced the city police to let the smirking Russian off with a warning. A final indignity: The police captain, who from time to time made himself available to the agency, extracted a small vengeance for being pulled from his warm bed. He insisted that the security team dig deep into their wallets and hand over whatever they could scrape together as recompense to the barmaid for the indignity she'd suffered.

The next morning the report detailing Nosenko's eventful night on the town was front and center on Pete's desk when he arrived at his office. He bristled with dismay as he read the account. Yet by now he knew the all too frequent summaries of rowdy and offensive high jinks were only a distraction from the more profound problem that he'd come to live with in the wake of the Russian's arrival in America. After months of sitting across from the defector in the split-level house in northern Virginia where the interrogations were regularly taking place, Pete had come to realize with a new and vehement clarity that something was very wrong with Nosenko. And if he was right, with the agency, too.

THE INTERROGATOR'S ART, EXPERIENCE HAD taught Pete, is not so much about asking the right questions as about listening closely to the answers. Let the prey ramble on; and all the while, hidden behind a capacious silence, you follow along, locking everything away in your memory. In the meditative aftermath, the skill will be in unearthing the telltale nugget that'd been inadvertently buried in even the most strenuous verbal contortions.

With a sense of foreboding, Pete had assembled the mountain of Nosenko's recorded transcripts and began the arduous task of once more going back over them. He was looking for the nugget. His working thesis, one part a wild hunch, another part a veteran's shrewd instinct, was that a consequential mystery lay buried at the operational heart of the entire Nosenko affair.

As a first step, he catalogued the cavalcade of lies. It was a very tangled web that Nosenko had been weaving. Nearly all the defector said, Pete judged, was invention. And adding to Pete's ire, Nosenko, forever shameless, apparently felt no compunctions about mercurially changing his stories and then taking great umbrage when caught. After the initial weeks of interrogation, Pete's damning inventory included—

Nosenko's solemn account of his own life and career. When put under the microscope, little in this fanciful autobiography held together. Pete, still fuming years later, wrote from his desk in Brussels: "His stories of his marriages and divorces rang false, his military service made no sense, and his . . . entering the KGB clashed with the administrative requirements known to us from other sources."

Then there was Nosenko's prideful summary of his service in the KGB. He claimed to have had a high-flying career, rewarded with promotion after promotion as he moved up within the American Department of the Second Chief Directorate, the branch that kept a watchful eye on tourists and the Moscow embassy personnel. But what had Pete scratching his head was that while Nosenko was overseeing some of the KGB's most promising bits of skullduggery against American targets in Moscow, his operational duties were often cavalierly put on sudden hold. Instead, he's dispatched as a security babysitter on at least eight trips taken by an eclectic assortment of Soviet groups traveling abroad. There's Nosenko, allegedly the Second Department honcho, only he's abruptly heading off as a lowly security thug to keep an eye on Soviet boxing teams going to London and Cuba or tagging along with diplomatic delegations while conferences drag on for endless months in Geneva.

And as for his description of a heady climb up the KGB ladder, the timing of his promotions just didn't make sense. There he is, the newly appointed section chief of the Tourist Department, but only after a month in this important job he's back babysitting diplomats in Geneva. Yet no sooner does he return from the three months of banal guard duty in Switzerland than he's rewarded with another promotion. Then, a dull year later, despite his achieving no successes to flesh out his résumé, he's crowned the first deputy department chief. Which was confounding, but not nearly as inexplicable as his being sent off, just days after having been elevated to this powerful position, to Geneva once again on a monthslong assignment as a lowly babysitter.

Another concern: The more Pete worked through the transcripts, the more he grew convinced that Nosenko had no idea of how bureaucratic life played out in the corridors of Moscow Center. Under the interrogators' questioning, Nosenko had conceded that he had no knowledge about such mundane yet necessary activities as the procedures for sending cables or summoning files.

Then there'd been the troubling evening when Pete, not for the first time, was sitting hugger-mugger next to Nosenko in a Beltway bar at happy hour. A gaggle of off-duty secretaries (or so Pete guessed) swarmed in, and Nosenko appreciatively checked them out, making his inevitable lewd comments. But at the same time, a small, offhand question popped into Pete's always churning mind. "How about the KGB secretaries?" he asked. "Are they assigned to individual officers, or do they work in a pool serving everyone in the section?" Nosenko ignored the question. So Pete tried again, and this time he made it clear he wanted an answer. And once again, Nosenko refused to respond. Which meant: Either Nosenko didn't want to divulge such an inconsequential scrap of information—or he had no idea because he'd never worked in Moscow Center as he'd claimed.

It all made no sense. Even by the devious arithmetic of the spy business, Nosenko didn't add up to who he said he was. Such a disjointed career, Pete knew, was just not possible in any professional intelligence service, neither in Moscow Center nor, for that matter, in Langley. And anyone who had done even a short stint in the KGB would have at least a cursory knowledge about how the offices functioned.

An unavoidable conclusion was staring Pete in the face: Nosenko was sharing a legend. His biography had been tailored out of yards of KGB-manufactured cloth to fit an image they wanted to present to the West. But why?

One possible answer, and a reasonable one at that, was that Nosenko had been dispatched to bolster Moscow's fervent assertion that they'd had absolutely nothing to do with Oswald. Pete could well imagine the uproar and the anxieties that had swept through the Kremlin when they discovered, as would the rest of the world in short course, that the assassin had not only lived in the Soviet Union for a while but also had married a Russian whose family had ties to the Communist hierarchy. Grief-stricken American generals and politicians start adding two and two, and how long before the two superpowers are at DEFCON 3, the launch codes dusted off, and forty or so million lives are hanging in the balance? It made sense that Nosenko had been sent to America to play a role in calming things down.

Only the script Nosenko was reciting from flimsy memory had been too hastily drafted by the KGB to be persuasive. Its plot was absurd. The more Nosenko trotted out this contrived scenario to his CIA and FBI interrogators, the more it unraveled. The list of contradictions grew alarmingly long.

One day he said that the complacent Soviets had never bothered to perform psychiatric tests on Oswald after he'd attempted suicide. The next he matter-of-factly recounted that he'd read the shrinks' evaluations.

Or he told the FBI that he had no idea how or when the Soviet embassy in Mexico City had passed on the incendiary information about Oswald's showing up to apply to return to Russia just two months before the assassination. The problem with that story, however, was that he'd already told the CIA he'd witnessed the arrival of the cable at headquarters and had gone on to recount the message in detail.

Then there was his initial assertion that he'd "thoroughly reviewed" what he had described as the KGB's single, skimpy file on Oswald. Later, however, he'd told interrogators that the Soviet secret intelligence service had assembled a hefty eight-volume compendium on Oswald prior to the assassination, and he had "glanced" at only the initial file folder.

Similarly, he maintained for months that the Russians had not paid any attention to Oswald when he lived in Minsk. At another interrogation session, though, he suddenly recalled that there were seven volumes of surveillance reports from Minsk alone, as well as a pile of transcripts of the conversations in Oswald's bugged apartment.

His information about Marina Prusakova, the Russian woman who became Oswald's wife and accompanied him back to America, changed, too. First he claimed the KGB had no interest in her at all. Next he said they'd started keeping tabs on her immediately after she met Oswald in March 1961.

And on and on. There were sufficient inconsistencies about crucial facts to cause any reasonable intelligence analyst to shout from the rooftops that Nosenko was deliberately covering up the truth because, QED, his Moscow Center handlers had something to hide. And this gave Pete the willies. He didn't believe Russia had a role in the assassination. There was no compelling logic, nothing that could be gained by such a momentous act. It defied the pattern of

Cold War machinations. But at the same time he was convinced that in a ham-handed attempt to establish their innocence, the Soviets had recklessly muddied the waters. And as a result, dangerous consequences loomed.

Full of alarm, Pete decided to fend off the incipient disaster. Fearful that Nosenko's disingenuous tales "might have a mirror effect" and reflect a deliberate Russian cover-up, he hurried to the Warren Commission. On July 24, 1964, Pete appeared before a closed-door session and addressed the seven commission members. "Nosenko is a KGB plant," he declared, "and may be exposed as such after the commission report."

Chief Justice Earl Warren, the head of the commission, got the message: Nosenko and his wobbly stories were fool's gold. It would be a genuine risk to try to cash in on them to persuade the American public that Russia's hands were clean. In this way, disaster was averted. Pete's testimony had convinced the commission not to include any mention of Nosenko, or his specious exonerations of the KGB, in its final report.

But no sooner had Pete enjoyed this triumph than, still unsatisfied, he began wandering back through a chronology of the entire Nosenko affair. The Russian had walked into the safe house in Geneva more than a year *before* Kennedy had been shot. He already had been put in play by his Moscow Center handlers as a double, ostensibly a CIA agent in place embedded in the KGB. But *after* the assassination, a rattled Kremlin decided to co-opt Nosenko into their frenzied efforts to deflect blame. He'd be the perfect messenger to send a reliable insider's account to America. All they had to do was order him to come alive again. So a script was quickly written and Nosenko learned his new lines as best as he quickly could; then he was sent off to reappear in Geneva. Only this time, reciting the newly revised KGB talking points, he announced that he did in fact want to defect, and, by the way, he also happened to have firsthand answers to the great conundrum the Warren Commission was struggling to solve.

The only chink in Moscow's scheme was that Pete, for the past year or so, had seen through Nosenko's original overture. And those findings had allowed Pete, already suspicious, to proceed swiftly to demolish pretty much everything Nosenko had said about the Kennedy assassination.

But there remained one part of this multifaceted plot that Pete understood he still needed to work out. Why, he asked himself with a new awareness of the question's importance, had Nosenko shown up in the first place two years ago? What had Moscow Center been up to? What sort of clever operation had they originally hoped to set in motion? What was worth the risk of the KGB's sending an operative right into the hands of the enemy?

Pete grew wary of the answer he'd find. Yet he also suspected it would explain everything. It would be the revelation that would expose a heritage of treason.

Chapter 13

PETE DID NOT PROCEED QUICKLY. The solution could not be pulled—*presto!*—out of a hat. Instead, as he now recalled years later as he sat across the dinner table from two of the men who'd actively helped him to make sense of Moscow Center's long-running plot, it had been in many ways a collaboration. He might have led the charge, but he was not the only one trying to get to the bottom of things. Each of them had something to add, and while the whole might not have added up to the sum of its parts, it had been an invaluable partnership.

They were a fraternity bound by shared suspicions. Rocca, accompanied by the heavyweight duo of Angleton and Soviet Division chief Murphy, had sought out Anatoly Golitsyn's insider's take on the entire Nosenko affair. Their feeling was that the agency's prized KGB defector, the operative whose stories Nosenko had so improbably echoed, would have a canny understanding of the covert game Moscow Center was playing. The genuine defector, they believed, could help expose the counterfeit.

He did not disappoint. Golitsyn's informed appraisal, as captured by a hidden tape recorder, pulled no punches. "He is a provocateur, who is on a mission for the KGB," he declared, as haughty and self-assured as usual. "He was introduced to your agency as a double agent in Geneva in 1962," he added. "During all that time until now he has been fulfilling a KGB mission against your country."

Pete, for his part, had reached out with an increasing sense of desperation to Deriabin. As an SB Counterintelligence chief, it was

Pete's job to poke into dark corners. But as he continued to explore the Nosenko case, Pete had come to realize he'd entered into sinister territory. He needed to get confirmation that he was walking—stumbling was more like it, he often felt—down the right path.

Deriabin was not only made cognizant, as the jargon of their shared trade would have it, of Nosenko's defection, he was also drafted to listen to the recordings of the meetings in the Geneva safe house, those from 1962 as well as the more recent ones in 1964. Then, he was given the tapes of all the stateside debriefings that had been conducted in the northern Virginia safe house. He listened attentively to the meandering and too often contradictory conversations, the sound of Nosenko's snarling voice becoming a palpable irritant. Enraged, Deriabin annotated the transcriptions as if they were scholarly texts, filling the margins with a handwritten scrawl of corrections and critical comments. Yet his spleen was not fully vented, and so he proceeded to enumerate a long series of pointed questions he wanted the agency to pose.

Pete read the list and, impressed, decided there was no need for any CIA middlemen. Deriabin should be the interrogator. It would be a conversation between two KGB colleagues chewing the fat about life in their common workplace. They'd chat fluidly, without any of the all-too-often perplexed interjections from agency officers not familiar with KGB slang. Then again, the discussion just might turn out to be an indictment of a pretend spy who didn't have much of an understanding at all of how things worked in Moscow Center.

There were twelve conversations between the two Russians, each rambling on for at least two hours. At the start, Deriabin did his best to mute the temper stoked by his reading of the transcripts. He offered himself as a counselor, a friend who'd come to help a compatriot smooth his increasingly rocky relationship with the overly suspicious Americans. "I am here . . . in order to clear up all of these misunderstandings which have arisen and existed," he began.

As the days passed, however, and Nosenko offered up little more than a plaintive refrain of "I do not remember," Deriabin's belligerence grew. "Let us think about where the bones lie hidden," he challenged. But he couldn't get Nosenko, who continued to take refuge behind a fortress of professed ignorance, to budge.

With the end of the final exasperating session, Deriabin filed his evaluation. He concluded "that Nosenko did not enter the KGB when or how he said, did not hold the KGB positions he said he did, and did not handle Lee Harvey Oswald's file. The way Nosenko explained his presence in Geneva could not be true, and his descriptions of his education and military service were impossible in the real Soviet world."

All of which confirmed Pete's instincts, but did little to advance his search for a guiding operational logic. He still needed to get to the heart of the matter: Why had Moscow Center sent Nosenko in the first place? How did the would-be double agent fit into the larger game the enemy was playing? And as he searched for his adversary's motive, he at last began to understand the true significance of the abandoned pair of panties.

"Band box" surveillance, the watchers call it. Other than attaching an electronic tracker, it's as good as things can get. Teams of pavement artists box the target in, covertly keeping their collective eyes on the prey from all sides. And while it's effective, the risk is that with so many operatives thrown into the mission, the chance the target will catch on becomes greater. Nevertheless, the KGB took that gamble and went all out on John Abidian, a young American embassy security officer. That was the provocative story Nosenko had told Pete during their initial friendly conversation in the Geneva safe house back in 1962.

The KGB had decided on this full-court press, Nosenko had explained, after they'd realized whose post Abidian had come to Moscow to fill: he was replacing the CIA security officer at the embassy who'd been handling Pyotr Popov, the GRU major who'd been supplying the United States with a bounty of secrets before his arrest and subsequent execution. The hope was that Abidian would lead them, same as his predecessor had, "to catch another Popov." But Abidian never led them to any meets or dead drops. The most incriminating discovery, the Russian went on with a lascivious laugh, had been a pair of panties found beneath Abidian's bed.

Pete, who never liked to leave any loose ends, later checked with the bachelor CIA officer. Yes, Abidian explained without embarrassment, a pair of panties under the bed was quite possible back then. Just don't ask me to recall, he added, whose they might have been. Apparently there was a long list of possibilities.

With that small nod to due diligence, Pete had put Nosenko's story out of his mind.

Only now when he made his way once again through the transcripts of the sessions that had taken place two years later in northern Virginia, something jumped out: Nosenko told an entirely different story. In this new version, the KGB's surveillance of Abidian had been a tremendous success. In 1960, they had caught the CIA officer "setting up a dead drop" on Pushkin Street.

Why, Pete wondered, such a consequential change in Nosenko's stories? First go-round, all the KGB got for their conscientious efforts was a pair of panties. Yet in the new version, the opposition watchers had struck pay dirt—a CIA agent outfitting a drop site. A newly recalled memory could not possibly serve as a creditable explanation for such a dramatic and significant discrepancy. Nosenko had vividly remembered a superficial detail like the panties, but forgot about the dead drop? Absurd. In Pete's distrustful world, there could only be an operational motive for this total rewriting of events.

Certain he was on to something, Pete locked himself away with Abidian's case files and in short course made a series of telling findings. First, the agency did in fact have a dead drop on Pushkin Street. Only neither Abidian nor, for that matter, anyone from the agency had "set it up." Oleg Penkovsky, the Soviet military intelligence colonel who, until his arrest in October 1962, had been delivering secrets—"invaluable," "essential," and "critical" were some of the adjectives used in government reports to describe this intelligence—to both his British and American handlers, had handpicked this covert letter box. Its selection had been a shrewd bit of tradecraft: a dark and empty space tucked behind a bulky radiator in the lobby of an apartment building about a mile's stroll from the Kremlin. A matchbox loaded with microfilm cassettes could be easily wedged in place. And no one would go poking around unless, of course, they had a reason.

Next, Pete discovered that while Abidian hadn't established the drop, on one occasion he'd indeed gone to the Pushkin Street apartment house lobby letterbox to pick up a delivery. The file told the story in some detail, and even the workmanlike narrative couldn't disguise the taut drama of a secret agent working behind the lines on hostile turf.

THE PHONE IN THE MOSCOW apartment of the American assistant military attaché rings at about nine on the night of December 25, 1961, the operational case history began. When it's answered, the caller doesn't speak. Instead, he blows softly into the mouthpiece once, then again, and finally a third time before disconnecting. A long minute passes, and there's another call. Once again, the identical voiceless response: three deep exhales as if the caller were blowing out candles on a birthday cake.

This was the recognition signal that the attaché and his wife had been told to one day expect. As previously instructed, the attaché's wife alerts the embassy CIA station by sharing the prearranged password.

Paul Garbler, the CIA station chief, is tracked down at a festive Christmas party at Spaso House, the American ambassador's residence, and is tactfully summoned to the phone. He takes the call in the privacy of the ambassador's wood-paneled study and hears a jolly assistant announce that she's finally succeeded in tracking down the Christmas present Garbler wanted for his wife. Garbler, who'd never dream of sending anyone off on his holiday errands, patiently plays along, and within moments his attention is suddenly riveted by a single sentence the assistant has woven into the conversation. It's word code, but she might as well have broken out into the Hallelujah chorus: Penkovsky has signaled that he's made a new drop.

A lifetime in the field has taught Garbler the value of misdirection; operational security is a constant vigil. He controls all his instincts to grab Abidian, who'd also been enjoying the ambassador's Christmas hospitality that evening, and rush off at once into the frigid Moscow night. He returns, instead, to the party, and plants himself by the bar. The last thing he wants is for some perceptive diplomat, either friend or foe, scattered among the revelers to start wondering why the CIA station chief had raced out from his own ambassador's holiday shindig immediately after being called to the phone.

Positioned by the bar, Garbler calls out for another martini in a stentorian voice that's a bit too loud and assertive for a diplomatic gathering. And only if someone is watching him very closely would they see that he casually deposited the barely sipped drink on an end table before barking out his demand for another.

You'll have to excuse me, Garbler finally informs the ambassador, flashing a merry smile as he speaks. Think I've had too much holiday cheer. John Abidian, he explains, has kindly offered to drive me home.

The two CIA men go off into the night, but on the way to the car Garbler makes sure to weave drunkenly for the inevitable audience of KGB watchers. And while en route, Abidian is forced to pull over to the side of the road several times so that his boss can vomit explosively into the street. The undoubtedly amused KGB surveillance team can't miss the frantic detours. Still, there's little chance they notice what the two American spies spot when the car, as if by random, pulls for still another emergency pit stop directly in front of telephone pole 35 on Kutuzovsky Prospekt: There's an X chalked on the pole. It's one more confirmation: Penkovsky has made a delivery to the dead drop.

The next day it's Abidian's turn for a starring role, and he leads the Moscow Center surveillance team that is his constant audience over a typically banal route. Routine, any fieldman knows, stitches together the fabric of cover. He visits, as he does every ten days or so, the same barbershop. He gets his usual trim, and, as always, the watchers, stoic and blank-faced, wait in their car parked outside.

Afterward, more routine, he strolls to the nearby bookstore. It's his habit to take his time, grabbing book after book and thumbing through each volume with thoughtful attention. As always, the KGB men remain in their warm car; the bookstore holds no interest for them. And so they don't see Abidian hurrying through a back door and exiting on to Pushkin Street. A moment later he's in the apartment lobby, his hand thrusting behind the radiator.

There's nothing there. He can't find the matchbox. There had been the two prearranged signals announcing the drop. But the letterbox is empty.

Confused, disappointed, and, as any handler would be, suddenly concerned about the safety of his Joe, Abidian hurries back to the bookstore, entering through the back door. When he's finally done browsing, the KGB car is still planted outside, one of the watchers complacently slouching against the hood as he smokes a cigarette.

At the time, the entire incident had been filed away as an aborted operation, and a perplexing one at that. The questions it should have raised were pushed aside as other deliveries were successfully made. And the new product was spectacular; the microfilmed documents detailing the missiles the Soviets were installing in Cuba were so significant that Robert Kennedy, who had been one of the key decision makers during the Cuban Missile Crisis, marveled that they "justified every penny spent on the whole CIA throughout its entire existence."

But now looking back at the Penkovsky case, his thoughts spurred on by Nosenko's conflicting stories, Pete began to see the events in a new light.

He wondered: If Penkovsky didn't initiate the signals announcing a drop, then who did? The obvious answer was the KGB. But how could they know? A quick dive back into the files showed that the Soviets claimed at the double agent's trial they had not learned about the Pushkin Street dead drop until *after* Penkovsky's arrest in October 1962, nearly a year later than Abidian's futile visit to the location.

Further, if the KGB watchers hadn't followed Abidian out the back door of the bookstore—and the CIA security officer was enough of a pro to have noticed anyone on his tail—then how did they see him going into the apartment house on Pushkin Street?

There was only one possible explanation: A team had already been in place inside the apartment house, covertly positioned to keep an eye on the lobby radiator. Only the timeline didn't work. Penkovsky wouldn't be arrested for months, and during the period before he was grabbed he'd continued to make deliveries. He'd remained operational. So how did Moscow Center know about the drop site back in December 1961? Or, as Nosenko recently claimed, as far back as 1960? That chronology not only didn't work, Pete realized; it was impossible.

And still another mystery: Why had Nosenko mentioned John Abidian and the Pushkin Street drop in the first place? Why the two conflicting stories? For that matter, why had Nosenko in his initial account shared the gossip about the discovery of the panties?

What, Pete asked himself, was he missing? It was all there, he felt. Yet it hovered just beyond his grasp. It was the shadow that he couldn't identify.

All Pete could do, he realized in the end, was to return to the Nosenko transcripts.

BUT NOW HE HAD A mission. He didn't know what he was specifically hunting for in the many volumes holding the transcribed interviews, yet at the same time his confidence grew. He believed he now understood enough—his instincts, if nothing else, were primed—to spot a hidden gem he might have previously ignored.

It was a painstaking, often tedious journey, but in the end, to his great excitement, there it was. In fact, it had been there all along, since the very first conversations years earlier in the prim Geneva apartment with the narrow balcony overlooking the painted roofs of Old Town.

It was a time when Nosenko, eager to be helpful, had been regaling Pete with all the tricks "the first-class" KGB tech squad had put in play against the West. They'd succeeded, he bragged, in rigging ashtrays and vases with microphones that could clearly pick up conversations despite the background clamor of noisy restaurants.

"I remember one such instance," Nosenko went on as if casually recalling an arbitrary memory. "We taped the conversation of the American assistant naval attaché as he lunched in a Moscow restaurant with the Indonesian military attaché, Zepp."

"How do you spell that name?" Pete had asked.

Nosenko quickly obliged: "Z-e-p-p."

And then their conversation had moved on to other matters.

But now, nearly four years later, when Pete reread this small exchange it brought a sudden shock of recognition. Knowing what signs to look for, he quickly grabbed the Penkovsky file and began his search. It took some time to find the precise intelligence, but in the end his memory had proven faultless. It was all there as he'd remembered.

The story Pete read was this: British intelligence had been running Penkovsky along with the CIA, and MI6 had, on occasion, used a British businessman as a courier. When the entire operation fell apart, this legman, Greville Wynne, had been put on trial alongside Penkovsky and he'd received an eight-year sentence in May 1963. Wynne had been doing hard time in Vladimir prison north of Moscow when the Brits, suffering pangs of guilt, made a deal to spring him. And once he was safely back in England, the MI6 backbeaters had their go with Wynne, wringing from him whatever he could recall about his time with the Russians. Then they had obligingly shared the product with their American partners. And now Pete, his long chase finally coming nearer to its end, found himself unraveling a previously mystifying exchange in the pages.

After five difficult months in prison, Wynne tells his British inquisitors, he's abruptly brought back to Moscow for a new grilling. This time the thugs are focusing on whether, in addition to Penkovsky, he had met with other enemy spies during his frequent business trips to Eastern Europe. It had been a particularly rough time for Wynne. In between sessions, they're working him with no sleep, blaring electronic noise, and when all else failed, or perhaps simply when they got bored, a bit of muscle.

Then one morning he's brought before the headman, and there's a new tack. "Who is Zepp?" the KGB interrogator demands.

At this point, Wynne has been broken. He's beyond keeping secrets. But he has no idea what the KGB brute is talking about. I don't know, he answers truthfully.

"Tsk, tsk," the interrogator mutters with a remonstrative shake of his head. He flips a switch on the tape recorder in front of him. There's the clinking of tableware, indistinct background noises—a restaurant. Then there's Wynne's clipped British voice; it's loud and clear. And next Penkovsky's unmistakable voice, mellow, even lyrical, and he's distinctly asking, "And how is Zepp?"

The KGB man fixes Wynne with a triumphant glare as he abruptly stops the tape. Well? he challenges.

Pete was on high alert. He read with great care the explanation that Wynne, his memory refreshed by the recording, had shared.

The name, Wynne patiently explains, was "Zeph," not "Zepp." And it has no operational significance. Rather, it's the nickname that Penkovsky and he had fixed on an attractive bar girl they'd chatted up in a London nightclub. At that moment they'd just been two blokes sharing a sly nod and a wink about a pretty woman they'd recently met.

With that cleared up, the KGB interrogator nods sullenly and promptly moves on, eager to probe more productive arenas. But five years later as Pete read the exchange, he stayed with it. It held all his attention. And at that moment he suspected that the Penkovsky case was the connection that would explain Nosenko's mission to the West. It was the common link that tied Nosenko to Abidian, to Zepp, to the Pushkin Street letterbox, even to the panties.

Yet the precise relationship between these two operations, between Nosenko's appearance in Geneva and Moscow Center's arrest five months later of Penkovsky, still eluded him.

PETE WAS NOT SOMEONE WHO liked to think in straight lines since, he'd say, espionage plots were by their deceptive nature twisted affairs. Therefore, he began to backtrack, returning once again to Nosenko's conversations about Abidian. Think it through, he told himself.

And there it was: the dates.

It'd been staring him in the face all along. Nosenko, once in America, had said that the KGB watchers had caught Abidian setting up the drop in 1960. But the CIA agent had not entered the apartment house lobby on Pushkin Street until a year later, December 1961.

Nosenko, Pete now understood, had been deliberately attempting to convince the CIA that the Russians had no suspicions about Penkovsky until they'd spotted an embassy official setting up a letterbox before his arrest in October 1962. First they'd told Nosenko to share the yarn about their surveillance yielding nothing more incriminating than underwear. And while you're at it, they'd also instructed, throw out the name "Zepp" and see if it triggers anything. But when Nosenko was suddenly reactivated after the Kennedy assassination—as well as *after* Penkovsky's arrest—the owls at Moscow Center thought they might as well give the Russian another new story to peddle. One with a new false chronology. If the Americans would buy into it, it would suppress any suspicions that might have been raised after Penkovsky's sudden arrest.

Pete found further corroboration of this premise when he zeroed in on the date of the conversation between Wynne and Penkovsky that the KGB had ingeniously recorded in the hectic Moscow restaurant. It had to have taken place just a few weeks after Penkovsky had come to London and his relationship with Wynne had been finalized—and when the two had frolicked in the nightclub with Zeph. That would have been, Wynne told his British interrogators, the only time he and his agent would have discussed the woman they'd briefly met. It could only have occurred, he said with certainty, in May 1961.

May 1961. That was a full sixteen months before Penkovsky's arrest. There would have been no reason for the KGB to have set up an elaborate eavesdropping operation with concealed mikes in a Moscow restaurant back then. They weren't on to him.

Unless the KGB actually had been. Unless they already knew of Penkovsky's treason. That would also explain how they'd first learned about the Pushkin Street drop: they had been tailing Penkovsky, not his handler. They had known the location of the concealed letterbox *before* Abidian had searched behind the radiator.

Then why had the KGB allowed Penkovsky to remain free for an additional year and a half? He'd traveled to London and Paris, and he had access to secret government archives. He'd continued to use the Minox camera the CIA had provided to photograph hoards of classified documents.

This was the last hard knot that tied together the Penkovsky and the Nosenko affairs, and as Pete kept pulling at it, it finally untangled.

The Russians had let Penkovsky remain free, even to continue to spy, because there was a more important priority: the overriding need to guard the identity of the invaluable informant who had originally exposed the traitor to the KGB.

It was a source, Pete now understood, who, if the CIA had gotten an inkling of his existence, might have been swiftly tracked down. The Penkovsky operation was closely compartmentalized; only a handful of agency officers were cognizant. By a rigorous process of elimination, the agency could have identified the rotten apple.

Nosenko's first arrival in Geneva in 1962 had been part of the KGB's plan to conceal this prized asset. The interplay, when studied from this historical perspective, was obvious. Nosenko had been originally dispatched with specious stories and an invented chronology to prepare the agency to accept that Penkovsky, like Popov, had been caught by nothing more ominous than diligent KGB surveillance. It was a deception operation that had been set in motion well before the Kennedy assassination. The defector's original mission had been, to use the jargon, source protection.

Pete now saw it all: Penkovsky and Popov had both been betrayed by a high-ranking mole. And this traitor was still embedded in the very heart of the agency.

IN THE AFTERMATH OF THIS realization, the decision was made to send Nosenko off for a lulling vacation in the warm sun and sandy beaches of Hawaii. Yet as the defector cavorted in Waikiki with a red-haired prostitute, her services paid for by the generosity of the American taxpayer, Pete grappled with the dangerous implications of his newfound certainty.

Chapter 14

1964

"We have to do something about Nosenko," Pete said. His tone was controlled, that of someone merely offering a helpful suggestion. The many battles he'd waged on the fifth floor had taught him that anger and ultimatums only stirred the opposition.

He was speaking to Dave Murphy, the head of the Russian section, and Pete's immediate boss. They were friends, both old fieldmen; Murphy had served in Berlin during a volatile period when the West had tunneled under the opposition's communications network and the East's wall had gone up. There'd been a time when both of them had plenty to say about the Langley deskmen's knee-jerk cant. But standing in the shiny corner office with its panoramic view of the leafy suburban Virginia countryside, Pete suspected that Murphy had switched sides; it's difficult to make your way in the agency without becoming a bit of a politician, without spouting off about the Big Picture and using it as an excuse to do nothing.

So Pete, according to his memory of the conversation, prodded gently. "The debriefing is winding down. There's not much water left in the well. The time has come, Dave. We've got to decide what to do."

That seemed to upset Murphy. He suddenly looked tired, as if he hoped the problem would simply go away. At least that was how Pete interpreted the weary look on his boss's face.

So Pete retreated. "Of course," he suggested, "we might just walk away from this." With little conviction, he threw out a plan to resettle Nosenko, find him a job, and simply keep an eye on him. But as soon as he'd laid out that alternative, he couldn't help adding, "But it would cost us our chance to clarify what's behind all this—and that will come back to haunt us."

"Yes, certainly his story about Lee Harvey Oswald will," Murphy agreed, his firmness cheering Pete. Murphy might dread having to make a decision, but Pete now suspected they were in accord in their evaluation of the defector.

"The Oswald story is one big question," Pete began, growing bolder. "But there are others."

Murphy nodded slowly; it was as if the gesture required all his strength. "Yes," he continued, his tone drained. And in that same exhausted voice he enumerated all the questions raised by Nosenko's troubling account of the discovery of Penkovsky's dead drop.

Pete wisely chose to keep silent.

Murphy finally continued. "Maybe," he tried weakly, "he was just boasting? Pretending he did things he didn't? That he's a congenital liar, a con man?"

"Hell, Dave, we've been through this before. You know we've tested every one. They all collapse."

"How about all the KGB spies he's uncovered?" Murphy fought back. "Would the KGB sacrifice them?"

Pete swiftly demolished this argument. "Try to name one who was previously unknown and had access to secrets at the time Nosenko uncovered him. There aren't any."

"Okay," he surrendered. "So Nosenko didn't expose any active or valuable spies." A moment later, though, Murphy countered with surprising vehemence. "But a lot of people around here believe the KGB would never send out one of their own staff officers as a defector."

There'd been no plan for Nosenko to defect until the Kennedy assassination had changed everything, Pete pointed out.

Murphy reluctantly agreed, letting out a small sigh of professional sympathy as he contemplated the enemy's predicament. "Wouldn't the KGB guys who set this up hate to see their long-range operation go up in smoke just to pass Khrushchev's message to the Yanks?" he said. "I can feel their pain. But they sure couldn't say no."

Pete hesitated, wondering if the moment was right. Then he decided he might as well share the looming possibility that'd been troubling him. Once it was out in the open, Murphy would have to respond.

"After he's done his gig with the Warren Commission," Pete predicted, "he'll probably find some excuses to get mad at us and return to the USSR."

"At least we'd be rid of him," Murphy deflected, and Pete's heart sank. But quickly the facetious tone vanished. "Okay," he asked with resolve, "so how can we confront him? What's ever to stop him from walking out the door?"

All along both men had known it would come down to this. To this question. This decision.

Done equivocating, Pete charged forward. "The only possibility I see is to put him under guard while we put the tough questions to him. Depending on his answers, we should learn enough either to clear him—or to decide beyond question that he is a plant."

Once again Murphy took his time. Pete sensed the agony of a bureaucrat who's been backed into a corner and now had to make a difficult decision. "It boils down to this," Murphy said thoughtfully. "Either we hold and confront him, or we drop the whole thing and pretend to take Nosenko at face value."

Murphy rose from his chair and, his back turned to Pete, stood facing a row of windows. He stared out toward the Virginia treetops.

"Okay," he decided at last. "Let's try. I'll take the question to the director to see whether there's even a possibility of the first alternative."

Section 7 of the Central Intelligence Act of 1949 offers a small but quite effective loophole to the carefully restrictive US immigration statutes. The CIA director "in the interest of national security and the furtherance of the national intelligence mission"—objectives that are as broad as they are nebulous—can with a mere wave of his scepter and a token nod from the attorney general and the commissioner of immigration bring up to one hundred individuals a year permanently into the country.

There are, however, two caveats buried deep in the statute's fine print: The new arrival must be who he has claimed to be, and no foreign intelligence service can have orchestrated his defection. It's up to the CIA, and for all practical purposes no one else, to determine whether these criteria have been satisfied. And to give this admonition some bite, the defector, upon arrival in the United States, is not offered citizenship. Rather, his specific legal status is something more murky: "parolee."

Yuri Nosenko had arrived in America as a parolee. He would soon learn how tentative—as well as precarious—such an official condition can be.

But first Murphy, after having been dragged into action by Pete, went to CIA director Dick Helms and made their case. His argument, as he'd worked it out earlier with Pete, stuck closely to the legalistic thesis that the Russian was in brazen violation of the Section 7 standards. For one incriminating thing, Murphy had decided: "We have to certify that he is who he said he is, that he has real access to the information he gave, and that we believe the reasons he gave for defecting. I couldn't certify a single one of those points."

And for another: "That 'we have no information that any foreign intelligence service influenced his defection.' Hell, we're up to our ass in just such info."

Helms did some hemming and hawing, throwing possibilities and explanations for Nosenko's behavior back at Murphy, but he also listened as his Soviet Division chief parried them with assurance. In the end, the director was convinced; and the fact that he was an old buddy of both Pete's and Murphy's, that they had been in the Cold War trenches together, didn't hurt either.

On April 2, 1964, Helms, accompanied by both Murphy and the CIA's legal counsel for additional authority, offered a blow-by-blow recitation of Nosenko's mendacity to the attorney general, Nicholas Katzenbach. As the director's indictment neared the finish line, he pulled his punches a bit, only hinting obliquely at the larger machinations that might lie buried at the bottom of the Nosenko plot, but by that point it didn't really matter. The attorney general didn't want to get into a tussle with a trio of spooks, and anyway, this was their bailiwick, their defector.

Two days after a merry, unsuspecting Nosenko had returned from his government-sponsored romp in the lulling Hawaiian sun, a dark sedan pulled up at the Virginia split-level ranch that was his home. Two security officers, their easygoing manner and bright smiles all guile, informed the Russian that a routine polygraph test had been scheduled. Sure, Nosenko agreed; as a counterintelligence officer he knew this was a customary part of the debriefing process. Let's go.

About an hour later, Nosenko, flanked by the two security gorillas and a representative from his usual troupe of CIA inquisitors, drove up a long, twisting drive that weaved through a forest of tall trees. They were deep in the Maryland countryside, the breeze thick with bird sounds. A stately brick manor house stood at the end of the long approach.

The heavy front door flew open before anyone could ring the bell. Nosenko entered, and with unseemly speed his hostile interrogation began.

AT THAT TIME, THE HEAD of CIA's Polygraph Division kept a photograph of Allen Dulles, the veteran OSS hand who became the agency's first civilian director, on the wall in his first-floor office. It stared out at visitors, and across the headshot, Dulles, looking very self-satisfied, a fortress of Brahmin rectitude, had scrawled in his own hand, "Polygraph is our first line of defense."

The reality, however, was more nuanced. At best, it was a Maginot Line: it'd get the job done only if the prey charged blindly forward. A strategy of evasion, however, could defeat it. With shrewd coaching, a professional could learn to "beat" the machine. And when a spy's inner life was as dizzyingly complicated as Nosenko's—either he was a genuine double or he was a triple agent pretending to be a double—role-playing was second nature. Like a skilled actor, he can tell lies that are truths bolstering the role he's playing in the moment—and the bewildered machine will flash "tilt."

Pete's only consolation was that a session attached to "the box" was better than anything else he could think of. *Depending on his answers, we should learn enough to clear him—or decide beyond question that he is a plant.* At least that remained Pete's hope.

Straight-backed chairs stood on opposite sides of a long table, each seat facing in a different direction as if to reinforce that this would not be a convivial conversation. Adjacent, resting on a smaller table, was a Keeler Polygraph Pacesetter, the 6308 model; a formidable array of dials and gauges gave the 6308 the appearance of the instrument panel in the cockpit of a plane.

Nosenko sat down in his designated chair, and with perfunctory efficiency a blood pressure cuff was wrapped around his arm and a pair of finger electrodes were attached, one to each hand. Three separate channels would continuously monitor any changes in heart rate and blood pressure, as well as his breathing and whether he was starting to sweat ("skin resistance," the examiners called it); the reactions would be simultaneously recorded by a constantly flickering needle on a revolving roll of graph paper.

Next, giving Nosenko the first discomforting hint that this might not be an ordinary fluttering, the hair on his head was brushed back and an electrode was attached by a suction cup to his skull. He was told it would monitor his "brain waves." Only it didn't do this at all; it was pure theater. The electrode's sole purpose was to plant the notion in Nosenko's mind that he was trapped, that despite all his training there'd be no possibility of outsmarting the ingenious CIA machine.

The questions, with Pete's help, had all been devised to elicit simple yes or no answers. And in his wishful mind, Pete had imagined Nosenko with each terse response careening with an increasing sense of panic toward the inevitable disaster.

The examiner, an old pro from the Polygraph Division, had never met Nosenko before strapping him to the machine, and, deliberately, he'd been provided with only a superficial outline of the case. He read the questions without a hint of menace, his tone heavy with a bored, official necessity. Have you been sent to deceive the Americans? Were you sent by the KGB? Are you still under Soviet control? Did you tell us the truth about Lee Harvey Oswald?

When the questioning was completed, the examiner left the room. As he went over the results with Pete and members of the security team in the kitchen, Nosenko sat slumped in his chair. The finding was clear: Nosenko was lying. Yet the discussion dragged on for a while, and anyway Pete was in no hurry. The maxim is to let the target stew; his imagination will be jumping about in terror.

"I'm shocked and disappointed that you're not telling the truth," Pete, affecting surprise, announced when he entered the room.

Sprawled in his chair, Nosenko raised his head slightly, his gaze one of languid disinterest. It was only when the Russian noticed the three security hardmen who'd accompanied Pete that his mood seemed to change.

"I have never lied!" he shot back with the desperation of a condemned man.

"You have lied," Pete insisted. He did not raise his voice; there were no histrionics. Instead he played the imperious commissar, the obstinate official the Russian would've known only too well. "And this time you've put us on the spot," he continued in the same glum tone, as if explaining a traffic violation. "This test was an official requirement. Now your whole position in this country is in doubt. We'll have to go over these problems one by one. We'll stay here until we straighten them out."

"I'll prove that I've been telling only the truth," Nosenko promised. But the words lacked his usual swagger.

On Pete's signal, the guards approached. Undress, one of them ordered.

Nosenko was led in his underwear to a dark attic. The only furniture in the space was a metal bed that had been positioned in what appeared to be the middle of the room. It had been solidly bolted to the floor. Black paper had been taped across the one window, effectively preventing any natural light from entering. Army fatigues had been laid out across the bed, and Nosenko was ordered to put them on. When he was dressed, the guards hustled him downstairs.

Pete, now joined by a Russian-speaking SB officer, was seated at the long wooden table where the polygraph examination had taken place. Nosenko, a hangdog expression on his face, sat down, his chair facing his two interrogators.

Without preliminaries, Pete started in. It was to be the first session of many. In the grueling days and months that followed, Pete, often backed up by a rotating cast of inquisitors, tried to get explanations for all the contradictions, all the inconsistencies, all the impossibilities in Nosenko's stories.

Life in his attic cell, Nosenko would complain with reason, was "inhuman." There was no contact with the world beyond his dark room; no television, radio, books, or newspapers. For a while, he was given only subsistence rations: weak tea, a thin soup, a gruel-like porridge. He'd smoked incessantly since his fourteenth birthday, but now he was denied cigarettes. The summertime heat in the room, a tight space directly below the roof without either open windows or air-conditioning, was torture.

Yet the harsh treatment and the persistent interrogations, Pete conceded, "didn't break Nosenko." Beleaguered but still belligerent, Nosenko never confessed that he'd been dispatched by Moscow Center on a disinformation mission. He refused to provide the single piece of intelligence, the atom at the core of the entire affair, for which Pete had been so avidly searching.

One very plausible explanation for his steadfastness, Nosenko's champions would later insist, was an obvious one: despite some self-aggrandizing exaggerations about his firsthand knowledge of the Oswald case and his understandable lapses in memory, he was telling the truth. And Pete was chasing shadows.

Yet the Soviet Division didn't see things that way. They still harbored large and troubling doubts. And after nearly a year and a half of Nosenko's barbarous confinement in the attic, it was decided that things needed to get much worse.

THE TEN-THOUSAND-ACRE COMPOUND, WOODED AND isolated, stretched for miles along the banks of the York River in Virginia. A high chain-link fence wrapped around the entire site, and armed guards patrolled the periphery. If any hiker happened to stumble into the area, he'd be warned off by a circumspect sign: ARMED FORCES EXPERIMENTAL TRAINING ACTIVITY, DEPARTMENT OF DEFENSE, CAMP PEARY. Insiders, however, simply called the site "the Farm." It was the CIA's training academy, a school for spies.

Handcuffed, blindfolded, surrounded by a phalanx of hardmen, Nosenko was taken in a security van to the Farm on a boiling August morning in 1965. The agency's explanation for the move was that neighbors near the Maryland house where Nosenko had been imprisoned had grown suspicious of all the comings and goings at odd hours. And, in another wrinkle, the house and guards were getting to be a burden on the agency's budget. It would be more secure, as well as more economical, to build a facility to house him at the Farm.

A twelve-by-twelve windowless concrete room was erected deep in the Camp Peary woods. The CIA security specialists who designed the structure called it "a little house." Nosenko—and, in time, others—called it a "torture vault."

Gripped by fear, wondering if he'd ever be released, Nosenko was confined to this cell for the next two years. And the interrogations grew increasingly belligerent.

Chapter 15

Yet Pete kept waiting for the final act. He kept waiting for Nosenko to break—and to admit that he was the centerpiece of a carefully orchestrated Moscow Center plot. It would be the disclosure that would inevitably lead the way to the mole. It would point him to the traitor hidden in the agency who'd betrayed Popov and Penkovsky, the traitor who'd inevitably cost more lives, disrupt other valuable operations.

Communicating through a closed-circuit television channel that linked headquarters to the newly constructed prison at the Farm, Pete took charge of the interrogations. He was comfortable in his role as lead inquisitor; the hours he'd spent poring over transcripts of previous debriefings had given him a pedant's knowledgeable confidence, and after all, he'd been there at the start, in the Geneva safe house. At the same time, a trio of additional interrogators was stationed in the room with Nosenko, filling the lulls, waiting to pounce with a well-aimed barrage of questions if the Russian appeared to stumble.

But the prize Pete hunted eluded him. His frustration deepening, Pete nevertheless felt he'd no choice but to plow on. "Nosenko dug himself deeper with new contradictions," Pete, unapologetic, explained. "He clearly had not held the jobs he claimed. And he said he had participated in operations against the US embassy that he didn't take part in. His statements were designed to hide something."

Pete was not a vindictive man. But as he tried to break this one last important lock—*his statements were designed to hide something*—he grew reckless. Why? Well, he was his father's son, a patriot—and so much was at stake for the nation. In war—and every fieldman knew firsthand that the Cold War was active combat—ends justify means, and lofty concepts like right and wrong were sacrificed to necessity. Besides, it was his job, it was what was expected of him: professionals do the nasty things that must be done so that others can sleep soundly at night. Or perhaps, as his detractors would charge, as his quest continued, his motives grew more tangled. Yet whatever was driving him, there can be little doubt that in his zeal he resorted to the enemy's techniques.

Interrogators taunted Nosenko that he'd be locked away for the next decade. More than a year went by before Nosenko was allowed outside, and then only for thirty minutes of daily exercise in a walled pen, his view a sliver of sky. From time to time the polygraph sessions were deliberately rigged, Nosenko told he was "deceptive" regardless of the results; one examination left him wired and strapped to his chair for seven seemingly interminable hours, including the four-hour "lunch break" his interrogator had decided to take. Drugs were administered on seventeen occasions, and while the agency would adamantly contend that these were only necessary "prescription medications," Nosenko—and who could blame him at this despairing point?—suspected something more sinister. "I'm sure it was LSD," he'd charge. "I was simply floating. I was almost half-conscious and suddenly I couldn't breathe. I couldn't take air in. I could take air out. I almost died."

And as Nosenko's cruel internment dragged on, as pressure continued to be applied without achieving consequential results, the fallout inside the agency took the form of a long, drawn-out exchange of adversarial reports. It was, quite literally, a paper war.

On one side were the die-hard proponents of what became known as the "Master Plot," those officers who like Pete believed that the KGB had perpetrated a high-level penetration of the CIA and that Nosenko had arrived with a bounty of stories designed

by Moscow Center to dampen suspicions and throw investigators off the scent of ongoing betrayals. In staunch opposition were the officers who rallied against what they derided, in a rare display of institutional wit, as the "Monster Plot." This vocal contingent sneered that the mole was nothing more than a figment of paranoid imaginations, and doubts about Nosenko were inventively blown out of proportion to support the ponderous weight of a pumped-up and completely unreasonable theory.

In the internecine crossfire, Pete would ultimately become a victim. But whether his comeuppance on this occasion was that of the hunter or the hunted would remain a matter of perspective.

THEY CAME GUNNING FOR PETE. The opening salvo in this protracted and contentious battle was fired by an indignant group led by Leonard McCoy, a report officer in the Soviet Bloc Division. Concerned by the disturbing stories circulating through the agency about Nosenko's savage treatment, McCoy urged Murphy, his boss, to allow others in the SB to read the transcripts of the Nosenko interrogations. And a shivery Murphy, apparently feeling that the cold winds of institutional change were starting to blow, acquiesced.

Already up in arms over the length and brutality of Nosenko's internment, McCoy was predisposed to taking the Russian's explanations at face value. And when he was done making his way through the mountains of transcripts, he indignantly judged that Nosenko was what he'd claimed to be all along—the real thing, a bona fide defector.

That got the director's attention. Sixty days, Helms announced in a stern ultimatum issued in August 1966. That was all the time the SB had to wrap up its case against Nosenko.

Pete jumped at the chance. Here was his opportunity to put the interrogations into context, to connect—as McCoy's narrow textual analysis had failed to do—the dots, to show that when informed by detailed references to other parallel clandestine Soviet intelligence ops, all the pieces in this very intricate puzzle fitted together with a telling malignancy. With that grandiose ambition pushing him on—and also knowing that his four-year assignment to Langley would soon be coming to an end; the Brussels station chief's job was in the offing—Pete went to work. In an earlier life, he'd been a scholar who'd written a doctoral thesis, and once again he studiously digested thousands of pages, referenced myriad case files—the footnotes alone were in the hundreds—and earnestly drew intricate connections between Nosenko's statements and his own investigations into arcane Soviet operations. It would be his magnum opus: the definitive argument against Nosenko.

It concluded, as Pete would later summarize, "that a tightly compartmented section of the KGB had sent Nosenko to us as a provocateur." But even Pete, with a gracious evenhandedness, had to concede that his report "had *not*, with certitude, got at the truth that lay hidden behind his lies."

As things would perversely transpire, this conscientious report was the first giant step toward Pete's downfall. Full of enthusiasm for the challenge to set the record straight at last, he'd written 835 single-spaced pages (and, he'd dauntingly promised, there were two additional sections still to be written). However, what he had meant to be an exhaustive analysis was, others snickered, simply exhausting. He was branded a zealot, and his report was mocked as the "Thousand Pager." Even those in his camp, ardent supporters of the "Master Plot" hypothesis, thought Pete, at the very least, could use a good editor.

An SB team went through the report wielding a merciless green pencil and when they were done they'd whittled Pete's volume down to a more manageable 407 pages. In tribute to their colorful handiwork, the new version became known as the "Green Book." It would serve as the official SB position on Nosenko. It argued, same as Pete had, that Nosenko had been up to no good from the start.

Pete, however, was incensed. Like many an aggrieved writer, he raged that clumsy editors had killed all his darlings. His editors' sins? They had "lumped together many separate points of doubt" and, in the bungling process, "they transformed justifiable points of doubt into debatable (and unnecessary) conclusions."

But while Pete, with a true believer's orthodoxy, was heatedly taking on the infidels in his own camp, the Nosenko case was, in decorous bureaucratic stages, quietly being taken away from both the CI and SB Divisions. And not until it was too late did Pete realize that the world beneath his feet had given way, too.

At first, though, Pete suspected nothing. The director, his head starting to spin from all the conflicting analyses, decided to throw his newly appointed deputy Rufus Taylor into the fray. He was ordered to oversee the Nosenko case and push it toward a resolution. Taylor had previously been an admiral serving in naval intelligence, and with a wisdom honed by years of military infighting, he swiftly passed the buck to Gordon Stewart, an old CIA hand who, the admiral shrewdly recognized, had the added qualification of being one of Dick Helms's buddies.

Pete was cheered. From his self-interested vantage point, Stewart was the perfect choice: he'd a well-earned reputation for integrity (in fact, he'd soon be appointed the agency's inspector general) and he'd come to the Nosenko case cold, without any preconceived notions or theories about the KGB's machinations. With the hope of giving a helping hand, Pete gathered up the essential case files as well as his "Thousand Pager," added an explanatory table of contents, and then handed the entire thumping bundle over to Stewart.

Pete might have saved himself all the effort. Stewart took one look at Pete's oversized pile of documents—a thousand-page report! a mountain of case files!—and quickly reached for the Cliff Notes summary, the SB's edited Green Book. A reading of this abridged version, however, left too many questions unanswered,

Stewart perversely decided. It seemed, he judged, "unscholarly," "more like a prosecutor's brief" than an in-depth analysis. Which, in fact, was pretty much Pete's own take on the watered-down Green Book. But rather than go back to the original source material, Stewart decided he had had it with the SB Division. The time had come to put their investigation—or at least as it was outlined in the abridged Green Book—under the microscope. Stewart gave the job to Bruce Solie.

Solie, a member of the agency's Office of Security, was a long, lean, and terse midwesterner. With a cigar dangling from the corner of his mouth, Solie was in his quiet, no-nonsense way reminiscent of the lone gunfighter who rides into a besieged Wild West town to clean things up. And he was fast on the draw. Within only weeks, he shot the Green Book full of holes. He suggested that a new "untainted" inquisitor be brought in, someone less confrontational who'd handle Nosenko with "more objectivity." In fact, he had the perfect man for the task, and Helms and Taylor agreed. So Solie took the job.

OVER A NINE-MONTH STRETCH THAT began in October 1967, Solie reinterviewed Nosenko. His technique, as he explained it, was not to badger, not to trap the defector in inconsistencies, but rather to give Nosenko the opportunity to share his version of events.

When he was done, Solie, with a facile practicality, argued that the secrets Nosenko had disclosed to him were so valuable that Moscow Center would never have allowed these treasures to be shared with the enemy. These pieces of solid gold outweighed all of the defector's "inconsistencies" and "abnormalities." And he compiled some impressive numbers to support his conclusion. "Nosenko," according to a CIA review of the interrogation sessions, "provided identification of, or leads to, some 228 Americans and 200 foreign nationals in whom the KGB had varying degrees of interest, and against whom they had enjoyed varying degrees of success. He provided information on about 2,000 KGB staff officers and 300 Soviet national agents or contacts of the KGB."

"Cost accounting" was what the CIA insiders called this manner of analysis, and when they, too, added up the profit and loss columns in the Nosenko ledger, they agreed: Nosenko, despite all his flaws, was a good investment.

Pete, however, thought the books were cooked. And no less troubling, they were being doctored behind his back. If it hadn't been for Angleton, who conspiratorially slipped the Solie report to him, Pete would never have known of its existence. When Pete indignantly complained to his bosses, he was informed his "need to know" had expired when he went off to the Brussels Station. But Pete knew this was a hollow rationale, and an ominous one. First they cut off your access, and then they go for the jugular. The handwriting was on the fifth-floor walls. His future at the agency, he realized, was shaky.

But Pete refused to go down without a fight. Although now very much an outsider, he still managed to have a working knowledge of the interrogation sessions (he was, after all, a spy). With a snappish sarcasm, he recounted the new approach:

Solie: "Wouldn't you put it this way, Yuri?"

Nosenko: "Yup, yup."

Or—

Solie: "But you really meant to say it differently, didn't you?"

Nosenko: "Sure."

It had been a complacent, self-affirming technique. Pete, beyond concealing his disgust, fumed that it was "a whitewash," "a way to rationalize the doubts." The agency, Pete charged, desperately wanted "to be rid of the ugly implications that underlay the Nosenko affair—KGB penetration of the CIA." "The debate was decided," Pete recalled, "but not the truth."

Yet the agency was suddenly in a rush to move on. With remarkable speed, once-accepted facts were officially rebranded as paranoia. And Pete, too, was tarred by this same broad brush. Solie submitted his report on October 1, 1968. Three days later—a nanosecond in an institutional world where things normally proceed at a glacial pace—its conclusions were adopted.

Pete read Deputy Taylor's endorsement, and to his nose it all had the whiff of something that had been orchestrated in advance. It was as if the CIA, after years of observing how things were done in Moscow, had staged its own show trial. Taylor wrote:

"I am now convinced that there is no reason to conclude that Nosenko is other than what he has claimed to be, that he has not knowingly and willfully withheld information from us. . . . Thus, I conclude that Nosenko should be accepted as a bona fide defector."

And he was. Nosenko was released, resettled outside Washington, DC, and reimbursed $137,052 in lost pay. In addition, with the hope of mollifying any lingering hard feelings, Nosenko was soon put back on the payroll. He worked as a CIA counterintelligence consultant for about $35,000 annually and gave packed lectures brimming with stories of his time in Moscow Center to CIA and FBI personnel.

Yet Nosenko's restoration had also come at a high operational price for the agency. "A back alley knife fight," was how one astute observer described the acerbic, no-holds-barred feud between the Russian's accusers and his champions. "On one hand, the Soviet Bloc Division wants this, and it's in cahoots with the Counterintelligence Staff. On the other, the Counterintelligence Staff has long been leery of the Office of Security, these gumshoes that just run around sticking their feet in their mouth. So this element is at play in everything happening here." And Pete was one of the victims. The knife had been stuck deep in his back. His reputation, his career, his future at the agency—all had been wounded in the battle.

Yet he was not alone. For years to come, the CIA would be divided into two camps of warring belligerents. This knee-jerk antagonism would continue, growing more caustic, more personal with the passing decades. It would cause officers to disregard facts, to refuse even to contemplate completely plausible theories. To settle old scores, operational determinations would be made largely by internal allegiances, and methodical, objective analyses of events would become secondary. And this was a very dangerous way for an intelligence service to do business. "The current top brass are taking unnecessary chances to demonstrate contempt for their predecessors," warned William Safire, a *New York Times* columnist.

As for Solie, he was rewarded. He received the Intelligence Medal for his work in "resolving" the Nosenko affair. It went unmentioned that this was the same medal Pete had been awarded nearly four years earlier for building the case that Nosenko had been a KGB plant.

Chapter 16

1981

THIS, THEN, WAS THE LONG battle-scarred road Pete had traveled that had led him to the dinner table at an old friend's home on the suburban outskirts of Washington. And he was still not done. It did not matter that he'd been hampered, even rebuked, by those with whom he'd served. Despite being retired, he refused to be a has-been. Duty, principle, honor—these were notions that did not end with the formalities of employment or job titles. For Pete, the past would never be past—not while he continued to be convinced a traitor remained in place in the agency.

Like a man with a lingering disease, he had been living with his suspicions for so long that they had become an accepted part of his life. Still, he had kept sufficient perspective to chart their progression. He was certain: there was a direct line from the unexpected reappearance of a defector in Geneva to the mysteries surrounding the deaths of Trigon and Paisley. And the cure? As he'd walked the battlefield of Waterloo in the gray new light of day, it was as if the voices of departed warriors had risen up in the mist to coax him, the old Marine leatherneck, back into action.

But where to begin? Pete had decided to start his new hunt precisely where his last campaign had ended. He would start with Nosenko. The defector was at the heart of Moscow Center's plot. Pete understood the logic of intrigue. He told himself: "The KGB did not blindly launch its provocateur out into the blue in the vague hope that all would go well. It 'required' some way to keep track of his progress, give early warnings of his problems, and if possible help him out of those problems." And this agent runner—Nosenko's invaluable lifeline—had been burrowed deep in place inside the CIA. He was the mole.

Revitalized by his silent battlefield epiphany and armed with a vague plan, he'd left the comfort of his home in Brussels and returned to the hustle and uncertainties in America. It was this quest that had brought Pete to the dinner table of the man whom, back in another life, he'd concealed in a shipping crate and smuggled past the watchful eyes of the Russians into the West.

The wisdom of the trade is that professionals make their luck, and this axiom seemed to be holding for Pete this evening. He'd stumbled into the perfect cover. It had been Deriabin, after all, who had decided to play matchmaker and invited Pete's daughter, Christina, and Rocca's son, Gordon, a happenstance that gave Pete the breathing room to ease into his proposal. More good fortune, the Rock, another frontline veteran of the Nosenko wars as well as a "Master Plot" adherent, had happened at the last minute to be invited, too. The operational gods were certainly smiling on Pete that night. All that remained was for Pete to choose the moment when he'd make his pitch.

THERE IS NO TRANSCRIPT OF the dinner-table conversations, and no one in the room ever thought it would be beneficial to share their memories. Nevertheless, an informed version of how things would have played out that evening can be offered.

No doubt Pete would have waited until the two young spies were locked in their own private conversation. Perhaps he'd even waited until they had said their goodbyes and gone off together for a nightcap; it is undisputed that the sparks of an incipient romance had flashed from that very first encounter. Deriabin still had the talent spotter's gift.

As for his pitch, Pete's subsequent actions convey the substance of his thinking. He would have offered just the briefest of summaries of the Nosenko case, no more than the bare facts necessary; both Deriabin and Rocca had, after all, questioned the defector, spent time with him. Then, he would've pointed out that the timing was propitious to go back over old ground, even, if necessary, to reopen old wounds. The new president, Ronald Reagan, was a hard-liner, a man who looked at the Soviet Union without the rose-tinted glasses of détente. This president, he'd have explained, saw the "evil empire" for the enemy it was. And for the clincher, Pete would have stated that the newly appointed activist cloak-and-dagger CIA director, William Casey, was cut from the same militant cloth. With some force, Pete would've argued that now is the time: We have a new audience, and we can with good reason believe it will be a receptive one.

And after having laid the groundwork, he'd have made the move to enlist them into his crusade.

It was seventeen years after Nosenko's defection. It was nine years since Pete's retirement. But Pete revealed his intention to make his case one more time, to make one final attempt to convince the new leaders at the agency that Nosenko was not who he pretended to be. He needed both their help and guidance. And something more: He wanted Deriabin, whose insider knowledge of Moscow Center's machinations was still greatly valued by the SB Division, to write a supporting affidavit. Besides, it had been Deriabin who'd interviewed Nosenko twelve times, who had watched the closed-circuit television interrogations, who had transcribed the tapes of the earlier Geneva debriefs, and who had plodded through all the classified intersecting case histories. Who would be a better authority? This would be a last shot, one final do-or-die charge. And Pete, no doubt, would've made sure to share his conviction that nothing less than the future of America's secret intelligence services was at stake.

Then, having made his pitch, Pete waited.

He did not, the records show, have to wait long. After having been shunted to the sideline, this new promise of action, of being back in the secret center of important events, apparently was too exciting to refuse. The two old spies at the table agreed to help.

Rocca's role would be limited to consultations, to playing the wise backstage whisperer and sometime adviser to Pete. That had been his function at the agency, too; he offered historical perspective and didn't make the do-we-or-don't-we decisions that were the stuff of action-oriented operations.

Deriabin, however, agreed to jump directly into the fray. Pete had opened his heart to his old friend, and Deriabin would do nothing less in return.

FROM HIS HOME IN BRUSSELS, in March 1981, Pete sent his lengthy argument directly to William Casey. Its title pulled no punches: "Why Nosenko Is a Plant—and Why It Matters." And neither did the pages that followed. He went over old territory with a renewed vigor, hoping that new eyes would read his well-documented argument without preconceived prejudices. His concluding appeal, though, was more emotional, the years of frustration finally seeping through: "How and why could *so many* questions—even two or three of them—have arisen about any *genuine* defector? The questions alone suggest that Nosenko was hiding important KGB operations . . . and KGB penetration of the staff of the CIA."

Deriabin's argument was submitted as an appendix to Pete's longer report, and it was very much a reflection of its author—overbearing, belligerent, and impatient. It was a screed from a man who knew his own mind and had no time for fools.

"I have long wanted to make known my strong views on this case," he began with punch, "but I was warned to keep quiet because it might hurt my position with the CIA."

"Now retired, I still hope that CIA will change its wrong and dangerous position on Nosenko."

"I am certain the man known as Yuri Nosenko is a KGB plant."

Then having shot off his opening fusillade, Deriabin, with the methodicalness of a longtime counterintelligence analyst, went through "a small sampling of the hundreds of points of doubt raised by Nosenko in the eyes of a real KGB officer."

But it was his concluding words that both Deriabin and Pete hoped would scream like a warning siren in the new director's ears: "I became convinced—on the basis of the Nosenko and related case materials and my other experiences—that the KGB has penetration of the American intelligence services."

Director Casey received the two imploring reports and decided to respond—a bit. With a politician's instinctive evenhanded caution, rather than face head-on the threat that had been ticking destructively away inside the agency for years, he commissioned still another report. The job went to Jack Fieldhouse, who had come to the agency fresh out of Yale.

And for once, Pete had a game sparring partner. The Fieldhouse study was titled "An Examination of the Bagley Case Against Yuri Nosenko," and it was pretty much what it claimed to be. There were no diversionary dives into Nosenko's indefensible imprisonment, no attempts to excuse the obviously inflated and illogical statements Nosenko had shared about his firsthand knowledge of Lee Harvey Oswald's time in Russia. Rather, it was an attempt to put the case—as well as Pete himself—into historical context. Fieldhouse argued that the doubts about Nosenko were an unfortunate product of the Cold War times when they had been conceived—as was Bagley, too. For the officers who had grappled with Soviet hoods in the back alleys of Europe, it had been second nature to believe that the wily Russians were concocting one fiendish plot after another. Suspicion was the mindset of the day. And Nosenko, the Moscow Center bogeyman, was simply a self-fulfilling prophecy.

Pete considered this, and with an equable objectivity, he found himself acknowledging that Fieldhouse had a point: Those were indeed tense and hostile times. But if that was the best punch Fieldhouse could throw, well, Pete felt his argument could withstand the blow. Sure, he might be a relic, a man whose life had been lived in another more hostile era. Guilty as charged, he was ready to concede. However, Fieldhouse had not landed a single convincing punch against his pages of well-crafted arguments. The incriminating facts remained facts even if Fieldhouse chose to ignore them. If this was the best Fieldhouse could do, Pete remained certain that the case against Nosenko had walked away without a scratch.

But that didn't matter. The exercise itself was sufficient. Fieldhouse's diligence, however wrongheaded in its conclusions, had provided Casey with the rationale he needed. He stamped Pete's and Deriabin's reports "No Further Action Required." Then they were sent off to be entombed in the vast recesses of the agency's classified vaults.

Pete now had no choice but to concede that his defeat was both total and absolute. He had fired his last shot, and he'd missed. He had no doubts about what his sentence would be. His suspicions of Nosenko would be dismissed in the agency "as nothing but the product of potted preconceptions and wild theorizing by since-disgraced colleagues, incompetent and paranoid 'fundamentalists.'" And the mole would have slipped out of his grasp for good.

YET IN FAIRNESS, PETE'S MISSION to Washington to resurrect the Nosenko case cannot be deemed a complete operational failure. There was one ancillary, if unanticipated, success.

The house was big and white, and in its rear a meticulously manicured emerald-green lawn sloped down to a deep blue lake shimmering in the Virginia summer sun. For once the July afternoon heat was not a torture. For once the lakeside swarm of mosquitoes had seemed to vanish from the face of the earth. And arguably most miraculous of all, for once a spy's plot had come to fruition. For on that perfect summer's day at the genteel home of Pete's brother, retired admiral David Bagley, a former commander in chief of the

US Naval Forces in Europe, an old KGB's man scheming had borne fruit.

Christina Bagley and Gordon Rocca, barely more than two years after their meeting at the carefully staged dinner at Deriabin's, were married. The wedding of these two young intelligence officers, both of them children of retired CIA officials of once great renown yet now out of favor, brought a star-studded assembly of the agency old guard to the lakeside home. And for many of the guests, still ardent supporters of the Master Plot, it was a marriage that held the possibility that the torch was being passed to a new generation of true believers, and that old suspicions were not as dead as they might appear to be. James Jesus Angleton couldn't make it; he was off fishing in Canada. But he sent his blessings along with two impressively large salmon he'd personally hooked. The gift provoked some wonder until one wag pointed out that two was the precise number of fishes another Jesus had given to his disciples.

YET PETE HAD NO ILLUSIONS. And he did not have to wait long to find an additional reason to despair. It was shortly after the wedding that Nosenko stood on the stage of the dome-shaped auditorium adjacent to the CIA headquarters building and addressed a packed house. When he finished, the room full of agency personnel rose to their feet and gave the defector a standing ovation. With their applause, Nosenko's victory was complete—as was, by equal measure, Pete's defeat.

Part III

"It Takes a Mole to Catch a Mole"

1984–1987

Chapter 17

Brussels, 1984

At last Pete was ready to surrender. He had given it his all and had come up empty-handed. If there was anything he'd learned in his rambles across the plains of Waterloo, it was that there could be dignity in defeat, in a battle well fought yet in the end a dismal failure. And as for blame, like Napoleon he could pin it all on "rigid fate," on an intractable institutional mindset that would rather be pulled inside out by an enemy than confront the damage head-on.

For that matter, what had he really lost? By any measure he still had a rich life—a wife he loved, children he doted on, a world of eclectic interests whose pursuit brought him satisfaction. He could now, he told friends without a hint of gloom, draw a line, and at last walk away from his previous trade and never look back. He no longer had to save the world. He had done his bit, and now it was time for a new generation to rush unto the breach. Let Nosenko and his supporters have the last laugh. It wasn't his problem anymore. So what if they rose to their feet and applauded the rehabilitated defector? He couldn't hear the maddening thunder. At the start of each new day, a look out the window onto the convivial neighborhood bustle of Avenue de l'Orée reinforced his new, welcome reality: he was a long way from Langley, and even further from its problems and vendettas.

With the giddy intensity of an old man enjoying his second

spring, he reclaimed his life in Brussels. Everything about it, he told himself, and anyone who asked, pleased him. With excitement and not a hint of resignation, insisted Pete with a sincerity that struck many people who knew him as genuine, he looked forward to growing old comforted by his scholarly dabbler's life. In the future, his exhaustive research would be largely confined to identifying the varieties of trees and birds that filled the seemingly boundless medieval forest not far from his doorstep.

Only without warning Pete was ambushed. One morning just before Thanksgiving of 1984, a late November day when the glorious fall that had left the Bois de la Cambre animated with a blaze of colors had given way to the first icy intimations of winter, he found himself reading an article on the front page of the *International Herald Tribune.* He'd been having his morning *café crème*, gearing up for a day whose high point would be picking up the holiday *dindon* he'd ordered from the local *boucherie*, when his eyes locked onto the headline. "Ex-CIA Employee Held as a Czech Spy," shouted the piece that had first run in the *New York Times*. He read it once, and then again for the details, but before he'd even finished the second go-through it was as if he'd never left his desk in Langley, and he knew, as any old counterintelligence hand would know, that all he'd warned about had come to fruition.

He detected the rich promise of a larger story that had yet to be told. And as if in an instant, all his hard-won resolve, his rigid commitment to let the past remain a memory, was cast off.

Of the harried days that followed, as best as they can be reconstructed, the events moved forward without any orderly progression, and Pete's knowledge grew in small, disparate pieces. Like the blind man in the parable, he was a man reaching out to touch the elephant and making one small discovery at a time. How it all fitted together, what manner of beast lay out there, he still did not know. A further hindrance, his research was by necessity limited to private initiatives; there was no requisitioning of files from the Langley archives. Yet he was driven, a man with something to prove, and that counted for something.

And he still had his supporters. He had a network of colleagues and sources both in and out of the agency who did not raise their eyebrows with weary dismay at the first mention of the Master Plot. He discreetly reached out to Deriabin, to Rocca (now, of course, not just a former associate but his relative by marriage), to his old colleagues. (And it's not unlikely, although there's no hard evidence, that he probed the two recently married spies in his family; a bit of shop talk was, by many accounts, occasionally the main course at the holiday dinner table.) It wasn't long, too, before journalists caught the scent; oblivious of his inquiries, they were on the trail, digging into the same fertile turf.

The story that Pete assembled was, in its initial, still preliminary findings, this: The spy who had infiltrated the CIA was part of a long-running husband-and-wife duo, Karl and Hana Koecher. Karl Koecher had been first recruited by Czech intelligence and sent with his wife to America as, to use the jargon, a sleeper—an agent who slumbers away innocuously for years, all the time building his cover, until he's suddenly awakened and hurled into active duty. And the Czechs' long-term investment in Koecher paid off handsomely: He succeeded, miraculously, in worming his way into a vital division of the CIA, a repository bursting with top secret intel. Moscow Center ran Koecher for over a decade, shrewdly using the information Koecher delivered to blow one CIA covert op after another to smithereens.

Koecher, in fact, had provided—and this was the discovery that set Pete's mind truly racing—the intel that had doomed Alexander Ogorodnik, the double agent code-named Trigon.

Despite all the stories the KGB had eagerly spread, despite the confident conclusions of the CIA's own internal review, Trigon had not been caught by sloppy tradecraft, or the vigilance of the KGB's watchers, or simply the malicious intervention of bad luck.

It was now clear and indisputable: Trigon had been betrayed. By a mole.

Just as Pete had suspected all along.

It had been this steely hypothesis that had sent Pete back to America. It was the premise that had pushed him to once again make the case against Nosenko. It was the impetus for his recruiting Deriabin.

Yet Pete felt little sense of vindication. True, for the first time he had his proof that something had been very wrong inside the agency. At last he had support for his belief that a trio of invaluable operations—Popov, Penkovsky, Ogorodnik—had been betrayed. He no longer had any doubts that brave operatives had been slaughtered because of a mole's treachery. But he also understood that Koecher, like Nosenko, was only a small gear in the enemy's complex machinery of betrayal. Pete still remained, he wearily conceded, a long way off from identifying the traitor inside the agency who was their control, the master spy to whom they reported.

Yet Pete sensed that he was drawing closer.

As Pete felt his way forward, he began by trying to make connections between the Trigon and Nosenko cases. Of course, he'd attempted this before without any definitive results, but that previous exercise had been conducted before he'd known that Trigon had been betrayed by a mole.

With this new knowledge, a telling similarity came into focus. There was a consistency in Moscow Center's handwriting, in the tradecraft they'd employed in both cases.

The KGB had dispatched Nosenko with the cover story that he was a major in the Second Chief Directorate, its counterintelligence and security arm. It was a legend—and Pete, despite the vehement protests by Solie and the other naysayers, had long ago established, at least to his and Deriabin's satisfaction, that the defector's credentials existed only in the script the KGB had Nosenko memorize—

that served a shrewd purpose. From his first arrival in the Geneva safe house, Nosenko had boasted about "our first-class surveillance teams." He had volunteered that the diligent Second Directorate watchers had spotted an embassy attaché mail a letter to Popov. He had trotted out a gossipy yarn about the band box surveillance on another embassy officer, John Abidian, with the intention of planting the seed that Penkovsky, too, had been identified through the vigilance of the KGB.

Pete hadn't bought any of it. He'd been convinced that the intent of Nosenko's revelations had been to send the CIA scurrying off in wrong directions. And in ordering Nosenko to perpetuate this ruse, Moscow Center had inadvertently indicated the one secret it did not want revealed: the existence of a high-ranking mole in the agency.

Armed with this realization, when he returned to the Trigon case, Pete perceived an analogous attempt at deliberate deception. The KGB had once again disseminated the story that the vaunted Second Directorate watchers had gotten their man; they had noticed Ogorodnik's car parked in the same spot near Moscow's Victory Park on several occasions and perceptively recognized this must be a signal announcing a dead drop. To reinforce this version of events, a photograph of the head of the Second Chief Directorate arrest team receiving the Order of the Red Banner had been splashed across front pages throughout Russia.

Yet, this, too, was a fabricated tall tale starring the Second Chief Directorate surveillance teams. And it, too, had been elaborately designed to conceal the same big secret: the mole inside the CIA who'd caused the case to come crashing down.

Spurred on by this similarity, Pete went hunting for others. And these back-bearings bore fruit. For one thing, both cover stories had been carefully manufactured to throw a bit of smoke around precisely when the KGB had first tumbled onto their targets. The deliberate obfuscations raised the distinct possibility that Moscow Center for a period of time before the "official" arrests—years, perhaps—had been feeding the double agents false information to pass on to the Americans. Pete knew that if he were in the enemy's shoes, if he knew a double was working covertly in his backyard, he'd take advantage of the situation. He'd feed him disinformation that'd have the unsuspecting opposition charging off in wrong directions. A mailbox operation, they'd call it on the fifth floor. A scheme to send misleading messages.

And for this trick to be truly effective, it would require an agent in place in the CIA, someone who could tell his Moscow Center bosses precisely what the agency most wanted to know. Guided by this insider's knowledge, the KGB would feed the double carefully scripted packages of intel. There'd be just enough truth in the disclosures to get the excited agency analysts believing they were mining a gold seam. The bulk of the delivery, however, would've been entirely invented. And entirely misleading. It would be sure to send the CIA scurrying about futilely for a long, long time. *A wilderness of mirrors, Angleton had called it. Nothing was what it seemed to be.*

And now, despite his being very much an outsider in any tactical sense, Pete found a path to charge forward. A way, as they say in the trade, to shake the tree.

Before Pete had been a fieldman, he had been a burrower. As a graduate student, he'd spent months in the National Archives ferreting through boxes of government documents from a bygone era while researching his doctoral dissertation. It was in both his nature and in his training to look at things like a historian. As he began to contemplate an endgame, a historical perspective informed his quest to find the mole.

"Every traitor uncovered inside intelligence services, East or West, during the Cold War," Pete was firmly convinced after a lifetime of study, "was caught only because of tips from sources inside the enemy's camp." To prove his point, Pete could reel off a long list of specific cases, a roll call of traitors who'd been burned by insiders. "It takes a mole to catch a mole," he'd lecture.

It was an insight, Pete now recognized, that offered a strategy, and possibly even an operational advantage. For Pete had found his first mole: Karl Koecher. He would use him to catch another traitor: the mole Moscow Center had taken so many elaborate steps to protect.

With a newly restored confidence, Pete went into battle the next day.

Chapter 18

1962

IT WAS NOW A PAPER chase, and, a distinct disadvantage, Pete was confined to what would be derisively called in his former life "open sources." That is, what any civilian burrower can rustle up if he digs long and hard enough. Just make sure, a professional might sneer, your library card hasn't expired.

Yet on the plus side of the ledger, Pete now had the advantage of knowing what he was looking for. At the heart of his inquiry was his search for the connections in Karl Koecher's operational life that could point the way to his control. *It takes a mole to catch a mole*, he knew. True, Pete could not be certain that he'd recognize the signs when he first encountered them. He suspected it would be necessary to chart the entirety of the story, to browse slowly and speculatively, and then, once informed, look back with the advantage of hindsight to see how the pieces fit together. "The past pervades counterintelligence work," he'd lecture. And at this thorny juncture the chronology of events would also be a call to action. "The long look backwards," he called it, at once his mobilizing drumbeat and his strategy.

Or at least that was his hopeful plan. Back-bearings, as any counterintelligence analyst knows, can be a very tricky business. His search also had the added challenge of trying to uncover clues that a resourceful spy—an agent who'd managed to worm his way into the heart of the CIA—had taken assiduous care to conceal.

Still, Pete found reassurance for the systematic, painstaking approach on which he was embarking in something he'd heard during his first bright-eyed days in the field. Bill Hood, a dapper, high-living gentleman spy who had been a veteran of countless audacious ops both during and after the war, had been chief of the Vienna Station and had mentored the neophyte secret agent through, Pete would gratefully acknowledge, "the counterespionage operations that were to shape my career." One of Hood's constant refrains was, "You need to go forward before you go back." In truth, Pete wasn't quite sure what to make of this piece of advice when it had been first bestowed. But now he grasped its guiding wisdom: assemble all the facts before attempting to sort things out. You had to make your way to the end of the rainbow before you found the pot of gold. Or, for that matter, the mole.

With his ultimate mission firm in his mind, Pete plunged into the murky operational waters where Karl Koecher had swum undetected for nineteen years as a spy.

FRANZ KAFKA, PETE SOON DISCOVERED to his surprise, had helped pave the way for Koecher's entry into the covert world. It is the grim, gray 1960s in Prague, a doleful period of tyranny and suspicion, and in his focused mind Pete might as well be standing in Wenceslas Square. Stalin's wrathful ghost still haunts the Czech Communist Party, and repression is served along with the morning coffee. Yet as Pete now came to learn, on a July afternoon in 1962, the bright, restorative sun of opportunity had unexpectedly shone through all the gloom and lit up twenty-eight-year-old Karl Koecher's very problematic life.

Until that fateful day, Koecher's prospects had been on a downward spiral. And like so many of Prague's victims, he had the secret police to thank for this free fall. It didn't matter that he'd laid the diligent groundwork for a promising career. First, he'd achieved impressive grades while taking a grueling regimen of math and physics at Charles University. Then Koecher, to his own greater satisfaction and his father's seething rage, quixotically went off to study film at Prague's Academy of Performing Arts. With his honors degree in hand, he landed script and camera jobs on several movies, and filled in the time between productions with stints as a comedy writer on radio and as a reporter for state television.

But all the while the domestic security arm of the StB, the Czech secret police, were tracking his every move and doing their diabolical best to make sure Koecher didn't stay in any job for too long. When he was a teenager, one of Koecher's classmates had shot a soldier. The friend was hanged, and through his association with this criminal, Koecher became "a person of interest." It was all downhill from there. At university, he was sentenced to three months in prison for "insulting a public official"—which was StB speak for kissing a girlfriend while a policeman happened to be standing nearby. After graduation, he received a "morality violation" for hosting a party at his apartment where supposedly underage girls were present; the two-year sentence was, at the last tense minute, suspended. And although he won a competition for a UNESCO job based in Cameroon, the Czech government harrumphed that he was "too politically dangerous to travel abroad." Demoralized, heading toward the crisis age of thirty, Koecher had come to realize the StB wouldn't stop turning the screws. Worse, he'd no doubt that it would be only a matter of time before they'd be permanently tightened.

But on that morning in 1962, Koecher showed up for his latest part-time job, and fate intervened. He'd been working as a guide for foreigners, and it was a position for which he was well suited; he spoke five languages with impressive fluency, possessed an insider's knowledge of Prague's cultural hot spots, and, if he remembered to disguise his contempt for the vulgar foreign businessmen and their vacuous wives who were his usual clients, he could summon up an appealing smile.

In the art deco lobby of the Alcron Hotel, he shook hands with his latest charge, George Kline, a professor of philosophy and Russian studies at Columbia (who, in the course of a long and venerable career, would become known for translating and bringing to the West the poems of Nobel laureate Joseph Brodsky). Is there any particular place you'd like to see, Professor? the young guide politely inquired.

Well, Kline began tentatively, I imagine there's not much chance you've ever heard of the writer Franz Kafka? It was still years before the liberating Prague Spring, a wary time when it was dangerous to mention the visionary Jewish writer's name, and reading his books was an offense that could lead to a stiff term of reeducation at a penal institution.

Koecher, however, not only had dared to read Kafka, he had something more to offer. He was dating Milena Jesenská's daughter, *the* Milena of Kafka's heart-wrenching "Letters to Milena," the muse who, a tormented Kafka exulted, was "like a knife with which I explore myself." And, oh, the feverish stories the mother had to tell! She had passed them on to her daughter, who, basking in the family's association with greatness, had shared them with her boyfriend. And now Koecher lavishly offered up these rare delicacies to the professor. He recited one incredible story after another as he led the way to Kafka's grave. Over a lavish dinner that night he continued the apparently inexhaustible reminiscences (albeit at this point probably embellished if not entirely invented). Before the charmed professor continued on to Russia, he promised he'd keep in touch with his new Czech friend.

Throughout this courtship the StB had been watching, keeping an eye on both the American professor and the troublesome Czech guide. And they'd been impressed. It no longer mattered that Koecher was politically dubious. That he was soft on revolution. Cavorted with intellectuals. The secret service talent spotters had a professional admiration for the ease with which Koecher had ingratiated himself with the American. He possessed instincts, which with training they could exploit. Join us, the foreign intelligence service generously offered Koecher. While in the next heavy breath they ominously warned, And save yourself from the internal security bloodhounds.

It was quickly settled. Like a character in Kafka, Koecher woke up one morning and found himself transformed into a spy. Nearly two years of intense instruction followed. Surveillance techniques, dead drops, beating the polygraph—the StB put him through a rigorous course. Then, to test what he'd learned, there were field missions, the most daring against West German targets. His tradecraft, his tutors rejoiced, was excellent.

His good fortune continued. As Koecher made his way in his new secret life, he was no longer alone. In 1963, he'd married nineteen-year-old Hana Pardemecova. She was a fun-loving, stylish blonde with saucer eyes, a pert nose, and a playful, hundred-watt smile. She attracted lots of looks and, she'd proudly boast, she always enjoyed the attention.

Koecher was another sort entirely. An StB evaluation not long after his marriage summed him as being everything Hana was not: "hypersensitive, hostile toward people, emotionally unstable, touchy, intolerant of authoritarianism."

Yet together they made a single perfect spy.

In 1965, Koecher was summoned to the apartment of his StB handler. You and your wife are being sent to America, the new spy was brusquely informed.

Koecher was taken by complete surprise. "What should I do there?" he asked, both bewildered and anxious.

"You are going to penetrate the CIA."

THE TWENTY-THIRD DIRECTORATE OF THE StB was the division that ran illegals. This was the trade's universal jargon for agents who floated through enemy territory without the protective net of diplomatic cover. If caught, they had no embassy card to get them out of jail free; they'd do hard time—if they weren't slammed up against a wall and summarily shot. It was an assignment that was played long, the passing years adding the patina of respectability to an agent's cover. Yet all the while the spy's standing in the shadows, patiently biding his time, waiting for the propitious moment to become operational. Pete knew from hard experience that illegals, the best of them as good as invisible to the opposition, could wreak havoc.

In 1965, the Twenty-Third Directorate, with the help of their allies in Moscow Center, began running the Koechers.

The couple arrived in New York in 1965 claiming to be political defectors, Karl moaning that he'd been fired from his job at Czech Radio because of his strident attacks on life lived under the boot heel of Communism. And before he'd set foot in America, he'd paved the way by reactivating his friendship with George Kline, the Kafka admirer he had won over while showing the sights in Prague.

Was Kline another talent spotter? Did he have covert ties to the sort of hush-hush folks in Washington who were looking to enter into practical friendships with Eastern Bloc dissidents? The answer to that question remained buried in files to which Pete no longer had access. But regardless of the mystery over who was ultimately pulling the strings, Pete had no trouble charting the blazing speed with which doors were flung open for Koecher, the anti-Communist immigrant.

A rent-free house was made available in Nyack, New York, and use of a car was thrown in for good measure. A job at Radio Free Europe was provided, and after a brief tuition-free spell at the University of Indiana, a larger grant materialized to study philosophy at Columbia. Hana, meanwhile, had landed a job sorting diamonds at the tony Harry Winston jewelers on Fifth Avenue.

But clearly, the records also showed, Karl took clever advantage of the opportunities that came his way. He buckled down at Columbia and, with brains and diligence, built an effective cover. He was a star in the university's International Relations Department, one of a handful of students personally selected by Zbigniew Brzezinski, soon to be President Carter's national security advisor, to be in his seminar. He earned his PhD in philosophy in 1971, and along with it came a coveted position in Brzezinski's Institute on Communist Affairs at Columbia. And when Koecher, who became a US citizen in 1972, picked up a CIA employment application at the university student activities office, it was Brzezinski who was only too glad to write the accompanying letter of recommendation.

Every prospective CIA employee was subjected to a series of three probing examinations, each a rigorous hurdle that had to be successfully negotiated before a job would be offered: a background check, a psychological evaluation, and a polygraph examination.

Karl Koecher passed them all with flying colors.

In November 1972, he was given a top secret clearance and started work at the CIA.

Chapter 19

THE DRAWING WAS DONE ON graph paper, the lines dark and bold, the lettering in a legible print, block letters and all capitals. It showed the layout of a Maryland shopping center. In small, neat boxes, the words clear and direct, was a roll call of the shops: Hamburger Hamlet, Giant Food Store, Sloane's Furniture Store, Ice Cream Parlor. At a glance, it had the appearance of a child's impromptu handiwork, something crafted while looking out the window of that family station wagon after Mom had dashed into a store.

Yet it was a map for a secret agent behind enemy lines.

And X marked the spot.

It was there, as bold as a beacon, just below the box marked "K-B Theater." A very carefully drawn X. As a further clue there was a bubble, the kind that appears above a character's head in a cartoon, only inside this one were the words "telephone booth."

Koecher had received this map from his wife. She'd been on her way to her job in New York's Midtown diamond district and had just exited a packed subway car. As she weaved through the morning rush-hour crowd spilling about the Times Square subway platform, her handler passed her a Dunhill cigarette pack. He didn't stop. He didn't even break stride. It was a perfect brush pass. And wedged inside the cigarette pack, folded fastidiously, was the map. When she went down to Washington to visit her husband that weekend, she handed it over.

Now it was his guide. It was the dead of night, well after midnight, a time when the shopping center was as dark and deserted as a ghost town and every stray noise seemed as loud as an explosion. Koecher carried a brown paper shopping bag, the sort checkout clerks fill with groceries. Inside the bag was another bag, but this one was plastic, and it was firmly bound with strips of masking tape.

Koecher found the telephone booth and entered. He dialed the number for the correct time because, as every professional knows, you must be able to account for every movement; you never know who's lurking about. He listened to the recording and hung up. Then he left, making sure to leave the shopping bag on the floor of the phone booth.

He drove back home to Washington and crawled into bed. Only more often than not, he'd later share, he'd be too wired to sleep after a drop. He'd reach for a book and try to read, but concentration would be impossible. His mind wouldn't stop racing, his thoughts careening every which way. He'd be filled with a restless, excited energy. All he could do, he'd explain, was pray he wouldn't look too fatigued when he showed up at his desk a few hours later at the CIA.

Pete, who in his youth had also known the solitary dread of the spy sneaking through the night, the throbbing fear that the enemy was monitoring your every move, read the accounts of Koecher's deliveries with a particular understanding. Yes, he'd been there, too, a soldier in the silent battles of a silent war. Those experiences would forever be a part of him. Yet even his vivid reminiscences were pushed aside by his rage—his temper brought to a boil by Koecher's audacity and, with an equal measure of fury, the slipshod institutional precautions that had allowed the enemy agent to roam unimpeded through a rich hunting ground.

On February 5, 1973, Koecher, his top secret security clearance testimony to his impeccable character, started working as a contract employee in an ugly concrete-and-glass monolith in Rosslyn, Virginia. As he drove across the Key Bridge, the building rose in the distance like a raised fist, another of the brutalist office complexes that had been thrown up around the Beltway to house Washington's burgeoning executive army. This tower, however, was a repository of secrets.

It was home to AE/Screen. This was the cryptonym for the CIA's Soviet Bloc Translation and Analysis Unit. Wiretaps, field-agent reports, informant debriefs, intercepted documents—whatever clandestine pots the agency was stirring throughout the world that involved Russian-speaking individuals, the raw material made its way to AE/Screen. The screen interpreters and analysts would translate the initial product into English, often attaching helpful summaries, and distribute the transcriptions to the operational teams and specialists at Langley.

Koecher had received a 5 on his Russian-language evaluation test, the highest possible grade, a rating that judged him as fluent as a native. And he'd studied math and physics at university; this familiarity with science gave him an advantage over many of the interpreters trying to make sense of documents heavy with arcane terms or phone intercepts that discussed, say, missile telemetry and triggering devices for nuclear weapons. Yet none of the most interesting raw material—the sort of provocative secrets the enemy was eager to know—made its way to Koecher's desk. Day after dull day, he'd sit with his earphones on, listening to tapes that had little operational significance, hearing commonplace information that might just as easily have been picked up from browsing an edition of *Pravda.*

A resourceful and supremely confident secret agent, Koecher decided to do something about his predicament. And he moved quickly. A mere three weeks after being assigned to the backwaters of AE/Screen, he shot off a high-handed complaint to his CIA bosses: "My present position is by no means one which would require a PhD. I am interested in intelligence work, and I want to stay with the agency and do a good piece of work. But I also think that it would be fair to let me do it in a position intellectually far more demanding than the one I have now."

His self-assured demand worked. He was given access to highly classified raw material. It included wiretaps from the Soviet embassy in Bogotá, the tantalizing gossip that first encouraged agency talent spotters to make their steam-bath pitch to Ogorodnik. And once the diplomat was up and running as Trigon, Koecher also had a look at the product the agent had left at his drops. One astounding delivery, for example, focused on Henry Kissinger and how his private conversations with the Soviet leadership steered their deliberations as they struggled to make arms control decisions.

When Koecher's contract at the Rosslyn site was up, he found new employment at another agency division offering bright opportunities for a hard-charging spy—the Office of Strategic Research (OSR). This was the hush-hush team that assessed Soviet Bloc military threats and capabilities as well as arms control measures and nuclear treaty verification. If the enemy wanted to get inside the agency's head, to know what American policy makers were thinking and why they thought it, the OSR was as good a place as any in Washington to get real-time insights into this strategic mindset. In the high-stakes poker game of nuclear missile negotiations, a spy in the OSR could whisper to the enemy what cards the other player was holding.

There was so much rich material for the taking, Pete read with a sickening sense of dismay, that even KGB head Yuri Andropov, a sinister figure who'd be more likely to order a throat slit than mutter a word of praise, marveled that Koecher's deliveries were "important and valuable." This assessment was further reinforced by Andropov's authorizing a $40,000 bonus to Koecher; and the spy, living his capitalist cover, used the money toward the down payment on a fancy co-op apartment on New York's Upper East Side.

Pete, as would any seasoned officer, yearned to know the full scope and detail of the secrets Koecher had passed on. But his resources were limited. There was a lot he'd never be able to uncover;

and, he also realized, there undoubtedly remained a good deal the agency would never know, too. At this stage of his investigation, he acknowledged, any exertions to find out more would be not just time-consuming but also doomed. It would be wiser to store the intelligence he'd already uncovered in the back of his mind, to wait until these tentative findings could be clearly fit into the larger puzzle he was methodically assembling.

Better, he decided, to move on to another aspect of Koecher's varied operational life. And as he explored, he could not help but feel that there were many versions of Karl Koecher, an assortment of characters that lay submerged, each bundled on top of another like the hidden figures in a Russian nesting doll.

THE FOUR TALL WHITE COLUMNS, two on either side of the bright red door with its shiny brass knocker, had been designed with the hope of evoking the genteel grandeur of an antebellum mansion. And while it was a huge house, a sprawling seven-bedroom affair with a circular drive that twisted up to the front steps, this was a shiny suburban Tara. There were a lot of preening McMansions set on half-acre lots going up in Fairfax Station, a woodsy, affluent commuter town about twenty miles south of Washington, and this one was cut from the standard upwardly mobile mold. Its pretensions were as big as the Ritz.

It had been up for rent for a while; its owners, like many people in government work, had been posted overseas for a year or two. Before shipping off, they gave the real estate agent handling the property one inviolable instruction: don't rent to a bachelor; single occupants were too unreliable, too slovenly in their habits. It took a while for the frustrated real estate agent to find a couple. But as things turned out, in the end there were lots of couples, and several threesomes, too. The suburban house with the white pillars became known in certain circles as the Virginia In-Place.

It was a sex club. And Karl and Hana Koecher knew it well.

They also frequented Capitol Couples, a group that met in a DC restaurant for dinner on Saturday evenings before adjourning to a nearby hotel suite to satisfy other appetites. And they were participants in the frolicking at the Rush River Lodge, a rustic retreat where mattresses were arranged edge-to-edge in front of a massive stone hearth to encourage conviviality. In New York, they'd make do with Plato's Retreat and the Hellfire. But Washington, Koecher recalled fondly, was "the sex capital of the world."

Hana, according to the cheeky articles Pete read with curious fascination in the *Washington Post* and elsewhere, was a crowd pleaser. Partners enthused that she was "strikingly beautiful," "ingratiating," and "incredibly orgasmic." As well as very sociable: "Hana loved having sex with three or four men on the double bed."

Karl, however, was a more withdrawn presence, chided the participants who'd been interviewed. Sometimes he'd get into the swing of things, but more often than not, they complained, he'd go off and find someone to whom he could talk.

Pete considered all this, and his initial tentative conclusion was that these activities had every mark of being carried out with the aim of burning, as the trade with decorous restraint refers to the cruel business of blackmail. Hana, he read, boasted that she met several CIA officials, Pentagon bigwigs, reporters, and a US senator at the parties, and had merrily bedded them all.

Were these liaisons driven by a spy's mission, an attempt to gather information? Was there something specific the couple had been hunting? Someone they were trying to meet? A target they'd hoped to blackmail? To what covert end were they engaging in all these energetic couplings? Or was sex sometimes just about sex? A sport and a pastime without ulterior motives? Something absurdly human rather than operational? Pete at this point in his research could draw no conclusion, and once again he felt he had little choice but to store these questions, too, in a corner of his mind. The answers, he could only hope, would come into play as he proceeded.

So Pete moved on, but what he now read came as no surprise. Jan Fila, an StB agent, defected, and he put the FBI bloodhounds on the Koechers' trail. *It takes a mole to catch a mole*—and here was one more confirmation.

The bureau followed the Koechers for nearly three years, bugging their apartment, their car, their phones. And in the dogged process, they'd intercepted not a single incriminating word. The Koechers were very good at the business of espionage.

Finally, though, when both the FBI and the CIA realized the Koechers were on to their surveillance and were planning to hightail it to Austria, the authorities pounced. The couple was imprisoned in 1984, Karl pleading guilty in a secret proceeding while Hana's more circumstantial case continued to make its way through the courts. But two years later the Russians, grateful and loyal, made a deal to get them back home. They were part of the East-West prisoner exchange that allowed Anatoly Shcharansky, the imprisoned Russian dissident, to be released and immigrate to Israel.

It was with a measure of bewilderment that Pete read the besotted newspaper description of the exchange. The reporter was as much a cheerleader as an eyewitness as the couple trudged through the heavy early-morning shadows streaking across the Glienicke Bridge and slipped from West Berlin to East Germany, and, in triumph, to freedom:

"With his moustache and fur-lined coat, Karl F. Koecher looked like nothing so much as a fox. His wife, Hana, wore a mink coat and high white mink hat. Blond and sexy, with incredibly large blue eyes, she looked like a movie star."

Heroes! Pete raged. Yet he vowed that the last laugh would be his. *It takes a mole to catch a mole. A maxim and a strategy.* And he moved forward, convinced that his quarry was growing closer.

Chapter 20

IN THE MONTHS THAT FOLLOWED his examination of Koecher, there were others Pete could have consulted. He could have turned to his family of spies—Rocca, Deriabin, even the newlyweds. But he didn't want their help; nor did he need their blessings. This had become a solitary quest. It was his own countervailing mission through the rampages of betrayal. It was about the people who had ruined the agency; and it was about those who, with their sneers and dismissive taunts, had conspired to try to ruin him. As Pete proceeded, he answered to no one except his own conscience. And as things worked out, that was all the motivation he needed.

For now, his investigation gathered speed, grew more purposeful, and Pete's thrust forward focused on the "holes." This was his own bit of word code, jargon he'd invented a lifetime ago after he'd veered away from the fieldman's frontline exploits and plunged into the more insular, meditative battles fought in counterintelligence. "Holes" was his shorthand for the missed opportunities—"The seemingly trivial anomalies," he'd explain—that previously had been overlooked. "Carelessness, oversight, or blunder"—those, Pete maintained, were the primary reasons provocative clues were ignored.

But there was also another, more deep-seated explanation. Pete's own victimization, his being tarred with words like *paranoid*, *fundamentalist*, and *fanatic*, had led him in time down a pained path toward this understanding. "Why," he'd started out naively wondering, "do professional intelligence officers, trained to expect such hoaxes and paying fulsome lip service to alertness, fall again and again into traps?" Why didn't they see through Nosenko, through the other ruses and deceptions?

After years of grinding away at this question, he had come to a disquieting answer: this failure was as good as preordained. A thoroughly ingrained institutional mindset had made it impossible for the agency chieftains to confront untenable conclusions. "Wishful thinking prevailed," he decided with consternation. This rosy mentality had led the CIA away from tough judgments, from interpretations "that bothered or threatened." With blinders tightly fixed, the agency headed inevitably toward self-deception.

But Pete, a man even at this late date still in search of completion, felt this was his opportunity to demolish that narrow, minacious outlook. His self-assured battle cry: "I had enough familiarity with Soviet counterintelligence practices to distinguish between real and unreal, enough detachment to call things as I saw them (independent of groupthink), and enough confidence to face the consequences of uncomfortable calls."

Pete knew what he had to do next. In the wood-paneled cocoon of his study, his separation so intense that the world beyond the two steps that led up to the closed door might not have even existed for all it beckoned, he went to work identifying the "holes" in the Koecher affair. Reading, annotating, cross-referencing. An analysis of years of deception would, he believed, lead him to a carefully concealed deceiver.

WHAT HAD BEEN MISSED? WHAT questions had not been asked? What connections had been ignored that linked Koecher to someone in the agency? That is, what were the "holes"?

And could they point the way to another mole?

When the case was illuminated in that glaring light, the record of official omissions produced a maddening inventory of concerns. The counterintelligence officer in Pete wanted to know—

How did Koecher obtain his top secret security clearance? Even a cursory background check of his life in Prague should have raised red flags, suggesting at the very least an association with the internal security forces. Could Koecher have walked away from his artsy pursuits and disappeared for two years—two years!—into an intensive StB training program and no one would've noticed? Shared a perplexed word with any agency burrowers? After all, dozens of Czech immigrants on the agency payroll with less problematic biographies had been unable to win such a dizzying level of clearance. Why had Koecher been completely exonerated?

And what about his lie detector test? Had Koecher simply possessed the innate cunning to beat the machine? Or had the questions failed to be sufficiently probing? Perhaps the results had been misread? That was one explanation the agency, red-faced and defensive, had later suggested. But was this alleged misinterpretation an accident, incompetence, or something more nefarious? Or another more troubling matter entirely that no one had dared to raise: had Koecher been rehearsed by someone familiar with agency practices, someone who knew how the interrogators routinely set about getting to the bottom of an émigré's life?

Then, there were his postings. First, there is the question of how Koecher managed to find himself tasked to the ripe vineyards of AE/Screen. No less mystifying, when this assignment shapes up as a dead end, when only banal tasks are coming his way, what does Koecher do? He throws a hissy fit—and right away he's given the keys to the kingdom. That sort of accommodation struck Pete as not the typical hard-nosed agency way of doing things.

Another surprise: when Koecher's time is up at AE/Screen, he doesn't have to linger in the overt world for long before he lands in another covert inner sanctum, another grove of secrets. At a momentous time when the two superpowers were negotiating precisely how many nukes they'll need to intimidate each other and yet still have sufficient firepower to wind up on top if push comes to shove, the Office of Strategic Research was the precise place a spy would want to be—and that's where Koecher lands. Pete had seen enough of the world to believe that while one mystery is a conundrum, two are a conspiracy.

Did Koecher have help in getting these plum jobs? A mentor at the agency to guide his career? Or maybe, Pete suspected, his familiar sense of foreboding rising up again, it was a handler? Sending an agent into the enemy's camp was a risky operation. It required help from within.

Which got Pete thinking: All those times the Koechers were dashing off to sex parties, were these interludes cover for meetings with their agency handler? Hana Koecher had bragged about the CIA officers who'd shown up at these shindigs. The agency, however, had been too prissy to follow up. Once again the security bloodhounds had covered their eyes rather than confront the literally naked truth. Whom might they have seen if they'd dared to look?

Further provocative territory: the intelligence that Koehler had sowed and the Russians had reaped. It, too, raised, Pete recognized, a roster of incendiary questions.

For starters, it was a matter of record that Koecher gave up Trigon to the Russians. But when? The agency's version held that Ogorodnik had been blown early in the winter of 1977; the alarm bells had started ringing when packages left for him at dead drops were not retrieved and the product he'd at last delivered was slipshod, his photography not up to his usual meticulous tradecraft.

But now, working backward, Pete saw that Ogorodnik had been recruited by the CIA in a Bogotá steam bath in 1973. In the months leading up to that pitch, Ogorodnik's phone had been bugged, his apartment monitored; the agency had been diligently gathering the compromising material documenting his reckless life, the stick that'd be brandished. And the telephone intercepts, the microphones, all the irrefutable evidence confirming that this Soviet diplomat had been targeted by the CIA in Bogotá would have made its way to AE/Screen for translation and transcription.

Which was where, as of February 1973, Koecher had been embedded. *Four years before the agency claimed the KGB had an inkling that Ogorodnik was a spy. Four years before the agency said Koecher had betrayed Trigon.*

Now Pete's mind was racing. He was energized by a new perception of the spreading pools of intrigue. Spurred on by his realization that Moscow Center might very well have been made aware of the identity of the double agent working against them years earlier than claimed in the agency's narrative, he returned to the case history.

He reviewed: Ogorodnik sensibly chooses the carrot when the CIA's pass is made and goes operational, reborn as Trigon. Only the new secret agent is a flop. Trigon manages to scoop up the chicken feed that's been scattered around the Bogotá embassy, but nothing worth all the hush-hush trouble. Then, Trigon is suddenly transferred back to Moscow in 1974, and he now has access to solid gold intel. It's the sort of high-level policy dispatches and eyewitness HUMINT (human intelligence) that satellite photos can't capture. This is product that only someone embedded deep in the belly of the beast can obtain. The CIA is elated.

But Pete has started to look at this chronology in an entirely new and startling way. And he found himself thinking, not for the first time in his long career, that in espionage, like life, there are no accidents. In rapid succession one disturbing explanation followed on the heels of another.

Okay, he tells himself, thanks to Koecher, the KGB knew from the very start, even before Ogorodnik formal recruitment, that he was a CIA target. Then, when he's up and running, it's Moscow Center that shrewdly decides to transfer the spy back to Russia and put in him in position where he'll have access to the sort of insider intel that will have Langley dancing for joy in the fifth-floor hallways. And why? Because it's disinformation. It's all been inventively scripted by the KGB. Sure, there's some good stuff, but only enough to get the agency excited. The rest is pure invention.

At the heart of this plot, Pete felt as his years of disparate theories coalesced with a resounding thud, was the mole who was already at work in the CIA. He'd pass on to Moscow Center the gaps in the agency's knowledge, as well as what they were most eager to learn. Shaped by these insights, the "documents" would be composed and then passed on to Trigon. Who'd leave them in a hollow log for his handler, closing the circle of deception.

Bolstered by this knowledge, Pete went on to untie another knot in the tangled cord the KGB had wrapped around Trigon. Pete was still very much an outsider, he only had the public reporting on the intel Trigon had provided, but it had been authoritatively leaked that the spy had delivered a cable sent in 1977 from Anatoly Dobrynin, the Soviet ambassador in Washington, to the Foreign Ministry in Moscow. The message revealed that Henry Kissinger, then a former US secretary of state and national security advisor, had given the enemy guidance on how to deal with the new Carter administration during the ongoing SALT II discussions on parity in nuclear weapons delivery systems.

Trying to stem the flood of enraged responses to Kissinger's "treason," his defenders had jumped in to suggest that the cable had been fabricated. Pete, however, had suspicions about the wider machinations that had led to the cable's invention. He knew, as any Langley insider from the Angleton era would know, that concerns about Kissinger's loyalty ran deep inside the agency; Angleton's files doggedly tracked "suspiciously long" personal conferences between the secretary of state and the Russian ambassador and noted with equal measures of indignation and innuendo that Kissinger refused to debrief the CIA on the discussions.

Only now from his newly scaled vantage point, Pete saw the entire ingenious operational life cycle of this scheme. Its genesis was the mole passing on the intelligence to his superiors that Langley had it in for Kissinger. Moscow then decides to fan the flames and cooks up a cable smearing Kissinger. It's passed on to Trigon, who includes it in the package he leaves for his CIA handlers. Along with the rest of this intel haul, it's delivered to the OSR, where Koecher duly translates the document before it winds up on desks in Langley where, completing the closed circle, it sends counterintelligence agents charging down blind alleys.

Yet no sooner had Pete worked this out in his mind than he had another staggering thought: What if, he wondered, Koecher, working hand in hand with the mole, had actually written the cable Moscow had passed on to Trigon? That way, he'd have received, evaluated, and distributed his own handiwork. It was the sort of amusing exercise that'd delight a mischief maker like Koecher.

But was that what had happened? I'll never know for sure, Pete despaired.

All these questions. All these what-ifs. All these "holes" that needed to be filled. And a mole who still had to be found.

Chapter 21

IN SUMMATION, PETE, STILL CLOISTERED in his study, began arguing in his mind. But no sooner had he started than he abruptly stopped. He had no suspects. Not a single name. Yet rallying, he told himself he'd succeeded in narrowing down the list of possibilities. He knew where to look.

He knew the mole must be either a serving officer or someone who worked closely with the SB Division. Only those agency employees would have knowledge of the actual identities of the Russian spies who had been betrayed.

He knew the mole would not be a field officer, someone who'd be routinely rotated from overseas assignment to assignment. He'd be a deskman based in Washington, someone always available to guide and protect the double agents he ran.

He knew the mole would have an operational involvement with the phony defectors the KGB dispatched, the walk-ins like Nosenko whose primary objective was to reinforce the credibility of Moscow's disinformation. The mole's access would allow him to be a continuing source of—and this was Pete's shorthand term for a brew of agency secrets and gossip—"timely information" that would freshly oil the wheels of deception and keep the Master Plot smoothly spinning along.

And he knew the mole was worth everything to Moscow Center. He was the prize beyond all others. He'd burrowed so close to clandestine CIA operations that Moscow feared he'd come under suspicion if the agency looked at its succession of failures with an unflinching eye. That was why Nosenko had been dispatched—to protect him, to throw counterintelligence investigators off the scent. That was why cover stories about the tenacious KGB Second Department watchers had been disseminated—to disguise the mole's behind-the-scenes role in Penkovsky's and Popov's arrests. His identity had to be guarded at all costs.

Pete thought about motives and opportunities. About truth and disinformation. About diversions and cover-ups. About the real and the unreal.

I have enough detachment to call things as I see them, he reiterated. *I have enough confidence to face the consequences of uncomfortable calls.*

He spent a long time thinking about all these things.

And at the end, he was back at the beginning. Back where he'd started and back where, he now suspected, he'd always known he'd end. He had returned to the singular mystery that had helped instigate his investigation, that had played such a forceful role in resurrecting dormant thoughts and buried slights, that had sent him off to wrestle with the past. He once again had a clear vision in his mind: a sleek sailing sloop drifting indolently in the Chesapeake Bay, a burst transmitter's high-pitched squawking, and a bloated, decomposed corpse wrapped in diver's belts.

Part IV

"In My Sights"

1987–1990

Chapter 22

I'M ON TO HIM. I'VE got him in my sights," Pete rejoiced to a friend who had arrived in Brussels just as he'd begun to focus on John Arthur Paisley. Still, the hunt would be laborious, another exhaustive paper chase. He would need to cross-reference the touchstones between Koecher and Paisley. He would try to establish the links, while at the same time he'd be scrutinizing the entirety of Paisley's career in this new—and possibly incriminating—light. The task was daunting. Yet it was no longer a directionless inquiry, a blind search where he'd be careening about frantically. He had a target, one with a human face, an operational history. He would identify the holes. And if he succeeded in filling them, it would, Pete solemnly hoped, prove his guiding hypothesis: *It takes a mole to catch a mole.*

It's easy enough to imagine it was these words, Pete's own heartfelt credo, that were pounding away in his mind, shoring up his resolve, as he entered Paisley's murky world.

HE WAS CALLING HIMSELF WILLIAM McClure, Pete recalled. That was the workname Paisley was using back in the late 1960s when their paths had first crossed. And it was into this shared operational history that Pete now dived.

It was not an arbitrary starting point. Pete was guided, as he'd been years earlier at Deriabin's matchmaking dinner, by his unshakable conviction that Nosenko "was not just another case." Nosenko was "the Rosetta stone," the dispatched defector who "was at the heart of everything." "The passage of time," he continued to have no doubt, "had not made old betrayals obsolete." And now as he set out, a previously discarded scene from one of Nosenko's many combative debriefings came alive in his thoughts. It was a piece in the puzzle that was John Paisley.

It had been a desperate time at the agency, as Pete knew only too well. Things were done that others, standing on the lofty moral ground afforded by hindsight, found cruel and difficult to condone. But even today Pete had no qualms; he had a soldier's harsh practicality about ends and means. He had no regrets.

To convey the stakes weighing on him during the many hostile interrogations at Camp Peary, Pete liked to tell a story that had been making its way through Washington at the time. Earl Warren, as the chief justice also later recounted in his memoir, had initially turned down President Johnson's request to chair the commission to investigate the Kennedy assassination. Therefore, Johnson, who had made a career of getting his way, had countered that there were "rumors floating around" driven by Oswald's mysterious time in Russia. "The gravity was such," Warren remembered the president warning, "that it might lead us into war, and if so it might be a nuclear war." And not leaving anything to the chief justice's imagination, the president added graphically "that he had just talked to Defense Secretary Robert McNamara, who had advised him that the first strike against us might cause the loss of forty million people." "If the situation's that serious," the chief justice decided, immediately overruling his own objection, "I will do it."

Pete had been ready to "do it," too. He'd been very much part of the frontline defense trying to prevent the catastrophe that Nosenko's tales might instigate. He'd needed to get to the bottom of what Nosenko knew—and what he didn't know—about the KGB's ties to the assassination of an American president. And the clock was ticking. "There was," he was being constantly reminded, "a special urgency to get at the truth of Nosenko's reports about Lee Harvey

Oswald because of the time limits imposed on the Warren Commission."

At the same time, Pete pursued a secondary, yet no less weighty, mission. "We had a mass of reasons," he was certain, "to believe that Nosenko was a KGB agent sent to harm the interests of this country." These reasons, too, were a call to duty. As a patriot and a professional, Pete was determined "to dig out the moles the KGB was hiding behind Nosenko's stories."

With so much riding on the outcome, would restraint, a policy of moderation, have been a virtue? Or would it have been a naive, even dangerous, surrender?

Dave Murphy, the head of the SB Division, had said the time had come to stop "allowing Nosenko to call the shots." Pete, no politician, put it more bluntly: It was vital "to break" him.

THE ARRANGEMENTS FOR THIS NEW round of interrogations that began in 1964 required, due to Pete's many concurrent daily responsibilities in the SB Division, some technological innovation. The Office of Security, however, managed to work it out so that Pete could, in effect, be in two places at once.

A closed-circuit television monitor was set up in a secure room just two floors below Pete's office at Langley, and, at the flick of a switch, Pete could see the image of Nosenko 150 miles away sitting in a straight-backed chair in the small, brightly lit room adjoining his cell. The Russian faced a barred door, and from time to time his eyes avidly scanned the four bare walls, searching for a way to escape from his confinement. At the start of each session, he looked weary, a man already pushed to the point of despair. But when Pete began talking, when his disembodied voice filled the interrogation room, Nosenko would gamely raise his head and do his best to engage. One official who had witnessed these long-distance sessions compared it to the old RCA Victor ad, the obedient hound cocking his head toward the sound of his master's voice.

As to Pete's interrogation style, it varied as the directness of the inquiries and his mood dictated. Invariably, though, he would start out coolly, carefully going through his well-prepared questions. But when Nosenko grew more and more flustered, as the inconsistencies and fabrications piled up and the entire tower of lies seemed about to topple over in disarray, Pete's pace would change. The questions shot out rapidly, and he grew more belligerent. He'd be shouting, cursing, pounding away at the same weak link hoping it would fall apart. Yet the best Pete could get for all his forceful efforts was for a beleaguered Nosenko to admit that he was "looking bad" even to himself.

It was a sort of madness. The Russian would throw up his hands in melancholy despair, or he'd hang his head like a condemned man climbing the gallows steps. He'd concede that he had no way to explain the many contradictions and factual errors that he'd stumbled into. Yet at the same time he'd play the innocent victim, as if the lies he'd spewed had no relation to the man who had said them.

In this way, Nosenko remained intransigent. He would not break. He would not admit that he'd been sent by Moscow Center to deceive and misinform. "Limited by morality and the law," Pete conceded with a heavy sigh of exasperation, "we were not able to get a confession."

The daily interrogations, however, did not end with Pete. Often he was just the opening act for the team of inquisitors the SB Division had assembled. It was, on Murphy's orders, a small group. "You are dealing with a potentially hostile guy who is liable to go back to the Soviet Union," he explained with the well-honed caution of a former fieldman, "and so you don't want to expose too many officers." As for the interrogators' qualifications, it was a mixed bag. Some knowledge of Russian was helpful; Nosenko's English was not good, and it would disintegrate further under stress (or, Pete suspected, simply when he wanted to disengage). However, Murphy later explained with rueful candor, "there just wasn't squads and squads of highly trained fluent Russian-speaking CI experienced interrogators." And apparently were there only a few officers available who had a deep knowledge of KGB operations and tradecraft. So the agency had to settle for vague, more subjective criteria when selecting the participants; hollow, gratuitous adjectives like "level-headed" and "tough-minded" were used to describe the team members. Pete, constrained by his own failure to break Nosenko, would try, for once, to be diplomatic. "Everybody has to get his experience somewhere," he'd say with valiant logic. "I think many officers I have known have done brilliant and complete interrogations without any prior experience." Yet, looking back at the way things had played out, it was difficult for Pete to avoid the judgment that by and large the team had offered a flawed response to a pressing and monumental opportunity.

But it was one particular interrogator's style and demeanor that vividly rose up in his memory as he now embarked on his new hunt. The picture Pete focused on in his mind's eye was that of a thin, hunched wisp of a man; even his beard was sparse and incomplete, as if he were still waiting for it to grow in. Yet despite his slight presence, there was something energetic, even strangely playful about him. He'd nearly bounce into the room, smiling and looking absurdly pert; he seemed to be enjoying himself immensely. He'd have a long list of questions, and, unblinking and methodical, he'd start to make his way through it. But once he detected a weakness, he'd pounce, tapping away at the same spot. Relentless and reproachful, he would not budge.

Pete, who had done this sort of thing before, knew there was a time-hallowed protocol for inquisitors. They would work in tandem, one offering himself as the sympathetic friend, the other as the intractable foe. From the onset Pete had, without a scintilla of guilt, presented himself as the bad cop. But this interrogator was apparently determined to play someone more cantankerous, a badder cop. Pete could still clearly hear the menace in his voice. He could still hear its taunting, booming cadence, a volume that seemed even more surprising considering the seeming fragility of the spare man making the threats. It would be a bitter and caustic assault. And it left no doubt that if he had his way, he'd string Nosenko up by the ankles till the lying stopped and straight answers started pouring out.

The officer used the name William McClure. His real name, Pete knew, was John Paisley.

But as Pete held on to that memory, as alive in his mind as if he were witnessing the scene today, he could not reconcile it with what he had recently learned. In the stunning aftermath of a former CIA man's body found floating in Chesapeake Bay, a gaping bullet hole in his skull, his corpse girdled in heavy diving belts, a *New York Times* reporter saw the prospects for an even bigger story. He poked around and discovered that Paisley and Nosenko had become friends, best buddies in fact.

Nosenko, after being vindicated and pocketing a hefty cash settlement from an apologetic CIA, had, along with his new American wife, settled into a pretty seaside community in North Carolina in 1970. And Paisley would often come calling. He'd sail the *Brillig* down to the Masonboro Boatyard and Marina, just a short hop from Nosenko's ranch house, and then spend time with the defector. "The visits were frequent and extended," the *Times* reported. Often, he'd hang out with Nosenko for a ten-day stretch.

Which seemed inconceivable to Pete. He still remembered the vehemence of Paisley's interrogation.

Unless it made total operational sense.

When the agency had exonerated Nosenko, putting him on the payroll and trotting him out as a wise man to lecture its officers, Pete had been astounded. It wasn't simply, as Pete had thundered, that "the concrete suspicions of Nosenko have never been resolved." The professional in him had risen up to declare it was dangerous. "It is irresponsible to expose clandestine personnel to this individual," he charged. How could the agency freely share the identities of its secret agents with the enemy? he wondered, dumbfounded and enraged.

But now he saw the consequences of the agency's "whitewash" as something more sinister. "During my years on the case," Pete felt assured, "Nosenko had no chance to make direct contact with Moscow or its representatives." "However, after the CIA put its trust in him, I presume he did manage contact with the KGB."

Pete had read lauding news reports that Nosenko had recently given the agency "valuable information" and exposed "new cases." But Pete didn't believe this for a moment. He knew the CIA had been conned. He had no doubt that "all this somehow benefited the KGB." Nosenko, Pete was certain, was once again leading the agency astray on Moscow Center's orders.

At the core of this long-running deception plot were the meetings between this active KGB agent and his visiting American handler. The very man whose growling animosity for the dispatched defector had once been on display for the entire SB Division to witness. Encounters that had provided, Pete noted with a twinge of professional admiration, bulletproof cover. And giving the scheme a further symmetry that undoubtedly brought smiles to the usually dour faces in Moscow Center, the two Soviet agents had been meeting in plain sight, in a seaside ranch house that had been paid for by the American taxpayer.

How else could the sudden reconciliation of Paisley, the relentless interrogator, and Nosenko, his aggrieved victim, be explained? The only logical conclusion was that both men had been carefully playing their parts from the outset. And that their recent meetings had been tactical, the interplay between a handler and his agent.

WITH HIS SUSPICIONS TAKING A more precise shape, Pete pushed on. It did not take him long to discover that Paisley, like Koecher, had been intimately involved in several other far-reaching defector cases. Yet as Pete now advanced, his hunter's eagerness was tainted by a sudden trepidation. It was almost as if he would've preferred to be proven wrong. For an unsettled moment, he shuddered at what he'd uncover as he probed this dark, sedulous conspiracy.

Chapter 23

A PITCHER RAN THE SAFE HOUSE. That was what Pete, who had been an athlete in own right, had learned with a good deal of interest on that rainy afternoon nearly thirty years ago. He'd been sitting in the front seat of the agency sedan making its way up the winding access road, the tires creaking over the carpet of wet leaves, on his way to another debriefing of Deriabin. It must've been sometime back in 1955, as best Pete could recall. But he vividly remembered the story the driver had shared.

Peter Sivess had jumped straight from the Dickinson College varsity to the Philadelphia Phillies. A blazing fastball had helped him set the school's strikeout record, and the scouts were touting the arrival of the next Dizzy Dean. Only the big-league hitters had no trouble timing his pitches, and the Phillies grew convinced they'd misjudged his talent. After a couple years, he started bouncing from team to team around the league. But then the war came, and he enlisted in the navy.

The navy offered Sivess the cushy opportunity to pitch for the service's team based in Quonset, Rhode Island, but he wouldn't have it. His parents, he informed the recruiters, were Russian-speaking Lithuanian immigrants, and he spoke Russian as if he'd grown up playing in the streets of Moscow, not South River, New Jersey. In wartime, that was a talent that could be put to better use than throwing a fastball, he argued. Besides, he added with a twinkle, my ERA doesn't sound too good in any language.

Sivess wound up working for the chief of Naval Operations as an intelligence officer. He had a string of cloak-and-dagger missions while stationed in the Black Sea port of Constanţa. When peace came, he stayed on in Romania as naval attaché, but that was just cover for his spy work. Then when the CIA was established, he was quickly signed on. Russian-speaking officers were coveted, and when you found one who seemed about the size and heft of the Washington Monument, that only made him more valuable.

Sivess was put to work debriefing Soviet defectors, and soon was promoted to chief of the Alien Staff. One of the chief's responsibilities is Ashford Farm, explained Pete's driver, wrapping up his story just as the agency sedan came to a halt.

The pitcher greeted Pete at the front door, a shotgun cradled in his massive forearm.

TODAY, EVEN AFTER ALL THESE years, the house still loomed large in Pete's memory. It was a big, rambling place, a gothic pseudo-Tudor monstrosity that the agency had bought for a song after the Pittsburgh steel magnate who'd built it in the 1920s went belly up, and then the subsequent owner stopped paying the mortgage. The redbrick house had seen better days when the agency took control in 1951, and an air of seedy neglect still prevailed after Sivess and his family had moved in. The furnishings were largely make-do, pieces found rummaging in the attic or picked up at local yard sales. The roof leaked, the green window frames were peeling, and the wallpaper in many of the eight bedrooms was water-stained, often tattered.

What Ashford Farm did have to offer—and this is what attracted the agency in the first place—was seclusion. The estate was tucked away on the edge of Maryland's Eastern Shore, surrounded by acres of thick woods that trailed down to the dark shoreline of the Choptank River. The nearest house was nowhere in sight. A chain-link fence surrounded the periphery. The access road was just a single lane, and it was a long, serpentine drive to the main house;

a visitor would be spotted before getting too far. Armed guards patrolled in precisely regulated shifts. The formidable Sivess, who liked to hunt deer and pheasant on the property, was on alert for other prey, too.

The house was a refuge for a rotating collection of the agency's prize defectors, enemy foreign agents who had switched sides. For many, it was their first introduction to America. They'd been hurried through immigration and brought straight to Ashford Farm. It would be their temporary home while they were poked and prodded by skeptical interrogators; and, as was often the wistful case, while they wrestled with the magnitude of what they had done, plagued by troubling thoughts of the loved ones they had permanently abandoned. In addition to Pyotr Deriabin, the guests from time to time included Anatoly Golitsyn and Nicholas Shadrin, as well as a squadron of other valuable CIA assets.

They were all names on a death list.

In November 1962, Pete recalled as sharp memories of the treacherous conflict he'd lived through as a young man rose up in his mind, the KGB chairman had approved a plan for "special actions."

This decree had fallen into the SB's hand, and it made for grim reading. "These traitors," it declared with glacial logic, "have given important state secrets to the opponent and caused great political damage to the USSR. They have been sentenced to death. In their absence, the sentence will be carried out abroad." The Thirteenth Directorate was instructed to train assassins who would carry out the death sentences anywhere in the world, including America.

In this secret war to the death, Ashford Farm was the CIA's inviolable sanctuary. It's perimeter, it assured the shaky defectors, could not be breached. The enemy, in fact, would never even learn where they were hiding. They were at last safe.

Putting all those memories aside for the moment, Pete returned to the present. From the seclusion of his own safe house in

Brussels, he made two phone calls. The first was to Deriabin, and the second was to Rocca. In the course of the two meandering conversations, both men confirmed, without Pete's needing to betray his motives or ask direct questions, that Paisley had been a frequent debriefer at Ashford Farm. Hiding behind his William McClure alias, Paisley had conducted many sessions with both Golitsyn and Shadrin. When the agency tucked Nosenko away for a short spell at the safe house, Paisley showed up, too; they spent hours walking in the woods, locked in deep conversation. Deriabin, for his part, had particularly strong recollections of his encounters with McClure, and for once the routinely condescending Russian had only good things to say about the intellect of his interrogator. They had played chess one afternoon, and Deriabin still took pride in his winning a hard-fought game.

In the contemplative days that followed this discovery, Pete reached two provisional conclusions. The first might be, he acknowledged, more guesswork than deduction; and, he further conceded, he had been driven to it by a raw anger that all the passing years had not succeeded in diminishing.

He had remembered: At the misguided core of the accusation that had tainted his career was the claim that the KGB had gotten hold of an actual debriefing of Deriabin. This stray bit of information—a "spiral," in the jargon of the CI analysts—had been cemented into place as the cornerstone of the rickety case that had identified Pete as a mole. The accusation had been that Pete had passed the memo to his KGB handler, and that it contained intelligence Pete had acquired during his initial conversations with the defector when the two had been holed up in the alpine safe house. But now a fuming Pete saw another, more likely explanation: it was Paisley's debriefing of Deriabin at Ashford Farm that had wound up on a desk in Moscow Center.

And hand in hand with that conjecture went another: the safe house had never been safe. And neither were the defectors. *It didn't matter that there was a fence around the chicken coop; the fox had a key to the front gate.* Ashford Farm—as well as a good deal of the secrets its inhabitants had been spilling—had been blown the moment Paisley took his seat in the cavernous, drafty salon that was used for interrogations.

How the hell did Paisley get this kind of access? Pete wondered. He was not an officer in the SB. He wasn't part of the CI squad. And he certainly wasn't a fieldman, at least not for our side. Yet he clearly had been able to infiltrate his way into many of the Russia Division's prized cases. With a mounting sense of dread, Pete could only wonder where else he'd find Paisley portentously hovering.

Yet even as he raised these speculations, Pete tamped them down. Later, he told himself, there would be time to explore Paisley's involvement in other operations, to zero in on how they might have been covertly manipulated. First, however, a single question, as complex as it was fundamental, needed to be addressed: Who was John Paisley? At this critical juncture, before venturing down other speculative roads, Pete decided he had to get a fuller measure of the man who was at the center of the puzzle. The man who had debriefed a roll call of Soviet defectors, who had strolled unimpeded throughout Ashford Farm. He needed to peer as best he could into another man's heart and see if it was a breeding ground for treason.

Chapter 24

1948

As he began scratching away at the genesis of Paisley's operational life, Pete did not have to go too deep before he found the unmistakable fingerprints of his old colleague, mentor, and friend, James Angleton. An assassination had brought them together.

The year was 1948, and, neither for the first nor the last time, Pete knew, turmoil was swirling through the Middle Eastern sands. Israel had just been born, and its delivery into the world of nations that May had been tumultuous. Five Arab armies invaded, and the new state fought back for its life. By June, however, the fledgling United Nations had managed to cobble together a rickety thirty-day cease-fire, and had sent a mission to the Middle East with the wildly optimistic goal of working out a permanent solution. The delegation was headed by the newly appointed UN peace mediator, Count Folke Bernadotte.

The count was a dashing, blue-blooded Swede who during the war had busily stirred many diplomatic pots: humanitarian activities that included a quixotic attempt to negotiate an armistice between Germany and the Allies, as well as Red Cross missions that, against all odds, had rescued thousands of POWs and Jews from Nazi internment camps. After a whirlwind tour of the region, stopping in Cairo, Beirut, Amman, and Tel Aviv, he confidently announced that he'd come up with a plan for setting things right.

The peace proposal did succeed in bringing about one manner of accord: both Arabs and Jews rejected it with mutual vehemence. But while its aspirational tenets had with each passing decade become increasingly impractical footnotes in an intractable dispute, Pete still found himself searching through dusty histories of the mission. Only now he focused his attention on the twenty-five-year-old radio operator who had accompanied the delegation.

It had been a hardscrabble journey that had brought Paisley into the inner circle of the Swedish nobleman, and Pete, as he pieced together the tale from news reports and books, couldn't help but admire the young man's tenacity and ambition. Paisley, along with his younger sister and sickly brother, had been brought up in the sweltering heat of small-town 1930s Phoenix, Arizona, and it had been a hand-to-mouth childhood. Their dad, a part-time labor organizer and a full-time drinker, had just drifted off, and it was left to Paisley's mother and his maternal grandparents to raise the kids. Clara Paisley worked as a cook in a tuberculosis sanitarium. The grandfather would leave the house in the predawn darkness with his lawn mower and rakes and head to the wealthy parts of town looking for yard work. The children grew up as outsiders, clothes threadbare and ill fitting, and all too often they'd walk to school in their bare feet since there was no money for new shoes. They had few friends, and their grandfather did his pious best to make sure their estrangement continued; dancing, popular music, card games, and trips to the movies were forbidden evils.

But young John had an active, inquisitive mind. He was just nine when he found his way to Vic's Radio Shop in downtown Phoenix, and Vic took a liking to the skinny kid who'd come by after school always asking questions; in this way, Pete suspected, he was the first of Paisley's many handlers. Vic helped him build a crystal radio set, and the boy would lie in his bed in the late-night darkness, earphone lodged in place, and listen to the exciting sounds of an active world beyond Phoenix and his dour, restricted life. It wasn't long before all Paisley could think about was escaping, about finding his way to what was going on out there.

When he graduated high school, Paisley saw his chance and grabbed it. He quickly enlisted in the Merchant Marines Maritime Training School at Gallup Island, across the country near Boston. He graduated the nine-month course as a lieutenant, junior grade, with a radio operator's certificate. And boats and radio, Pete noted, would remain central to Paisley's professional handwriting throughout his operational life.

When the war came, Paisley shipped off as a radio officer on voyages bringing much-needed lend-lease supplies to the besieged British and Russians. There were at least two voyages to the Russian port of Murmansk, where the Soviet Northern Fleet was based. The details of Paisley's time in the Soviet Union, at least as Pete could make out, were a mystery. Was the young radio operator affected by the courageous battle the ragtag Soviet forces had shown in the successful defense of the port from crack Finnish and German Arctic fighters? Was he approached as he made his way around the city by the NKVD who, Pete, the old SB hand, realized, made it their business to recruit sympathetic foreigners?

No less mystifying were the reports of how Paisley spent his time between voyages. He had managed, although the details that had led to the connection were never explained, to make contact with several professors on the Columbia University campus in upper Manhattan. Although he wasn't enrolled as a student, he was given reading lists and textbooks for a wide range of courses. Off at sea, he'd lie in his bunk at night making his way through the thick texts, just as, in a previous life, he'd listened to his crystal radio. In this self-taught way, Paisley, by the war's end, had mastered the basics of the Russian language. And why, Pete wondered, was Paisley suddenly so intent on speaking Russian?

There was no way of knowing for sure, but Pete couldn't help but suspect that the higher education of the Merchant Marine radio operator had been a consequence of his being groomed as an intelligence asset. Yet whether the helpful talent spotters were American or Russian operatives, that, too, was hidden away in files that were beyond Pete's grasp.

When the war ended, it appeared that whoever had Paisley in their sights had cut him loose. He drifted from job to job, a spell as a radio operator for the highway patrol in Phoenix, a hand on a steamship cruising to Alaska. Then Paisley enrolled in the University of Oregon—how he'd managed to scrape together the tuition and board fees was another puzzle—only to be expelled after a single term. Apparently, he'd not yet mastered the art of running a covert operation; he'd been discovered in his dorm room bed lying alongside an attractive blonde.

Paisley next popped up as a radio operator at the temporary headquarters of the newly convened United Nations in Lake Success, New York. Did Paisley simply answer a help-wanted ad? Was he encouraged, perhaps even directed, to seek out a position in the international organization? It would have been a rich hunting ground for any intelligence service. Again, the answers were unknown.

But it is certain that when Count Bernadotte's peace mission headed to the Middle East, Paisley, just twenty-five, was the group's radioman. His duties had him traveling through Israel, Egypt, Lebanon, Syria, and Jordan. And along the way, he met James Angleton, who, as it happened, was on his own secret mission.

A RECRUITER'S JOB, PETE KNEW from his own frontline experience, was one of seduction. The decision to enter the covert life, to endure the loneliness and uncertainties that accompany a life shaped by intrigue, required a commitment that transcended reasonable expectations. Not everyone was willing to sign on to the dangerous business of saving the world. And yet that was Angleton's enchanter's gift—the ability to inspire others to pledge their allegiance to the Great Game.

Today, Pete acknowledged, Angleton had become a disputed, even notorious figure in some precincts. And there was a strong argument for this appraisal; his flaunting of legal prohibitions and penchant for secretiveness, however patriotic his motives, had turned many on the fifth floor and elsewhere in the agency against him. But back then, having earned his spurs in wartime schemes in Britain and Italy, Angleton was the personification of the gallant, globe-trotting secret agent. His deep intellect, his iconoclastic quirkiness, his patrician demeanor—these were seen as the sort of transcendent qualities that would be needed in the dark, covert war against World Communism. New officers didn't just admire Angleton, they wanted to be like him.

His official position at the CIA in its early years was chief of Staff A. This boiled down to his running one group trying to steal the enemy's secrets as well another team hell-bent on preventing the opposition from stealing America's. And while carrying out this broad mission, which grew even broader as he kept stretching his mandate, he headed off to the Middle East during the uncertain fall after Israel had formally declared itself a state.

Angleton was looking for friends, and by friends he meant sources of information.

Initially, he'd arrived wary of the Israelis. The Soviet Union had been the first nation to grant the Jewish state recognition, and he suspected its support went hand in hand with a carefully entrenched intelligence operation; Moscow Center would infiltrate its agents into Israel, and then the operatives, wrapped in new cover identities, would disappear into the West. But when he met the shrewd Israeli spymasters, battle-tested veterans of intricate plots against both the British and the Arabs, running the country's maze of intelligence units—the Mossad, the state's centralized foreign intelligence agency, would not be formed until April 1951—he recognized kindred spirits. Impressed and intrigued, Angleton decided that he could work with the Jews. And the Israelis, with a corresponding practicality, decided they could work with him. A deal was tacitly negotiated: in exchange for the agency's help from time to time, Israel would provide access to the well-placed sources they were running behind the Iron Curtain. On this first trip to the new state, Angleton laid the groundwork for what became known in the agency as "the Jewish account," and for the next two productive decades, Angleton was its imperious manager.

Yet even as he forged this mutually beneficial long-term intraservice alliance, Angleton, barnstorming through the region, was also busily engaged in building a private army of his own inside the agency. He was on the lookout for operatives who could give the emerging CIA a discreet presence around the world. But there was one catch—they'd report directly to him. They'd be *his* spies.

Years later, one of the secret recruits from that freewheeling period shared Angleton's pitch. "Would you like to work with me?" he'd inquired matter-of-factly. "Not for the CIA. Just for me. I want you to be my eyes and ears, go on special assignments, stay clear of embassies . . . let things come your way naturally." The pay was five hundred dollars a month, and the expense account was bountiful.

Angleton's approach to Paisley, Pete theorized, would've been finessed with similar soft-pedaling. And, Pete had little doubt, it would have been preceded by a perceptively nurtured courtship. Angleton, as always, would have done his homework; he'd have gotten a good sense of his target before jumping in. Pete imagined the wily spy sensing the desperation in the young man who had grown up on the fringes. He'd have conveyed to him the promise of kinship, the opportunity of membership in an elite corps that was taking part in a great struggle against an evil enemy. And it was easy to see how Paisley, the self-made, even self-educated, man, would have been impressed by the aura of patrician polish that the confident, worldly secret agent radiated. The young radio operator who was so eager to leave his own past behind, Pete suspected, might very well have seen a way of reinventing himself, a possible future.

But while these thoughts, however well informed, were the products of conjecture, the timing of Angleton's pitch was a matter of record; the many press investigations in the aftermath of Paisley's disappearance were in agreement about that. And once again Pete could only applaud his friend's sly tradecraft. People, it stands to reason, were at their most vulnerable after they've witnessed a death.

The Jewish Sabbath would soon begin in Jerusalem, but despite its approach that Friday evening, the sun remained stubbornly high and there was no relief from the uncommonly warm early autumn weather. The unremitting heat was one thing people clearly remembered. The other was the peal of bells. For no liturgical reason anyone could offer, the bells had started ringing from somewhere deep inside the domed Greek Orthodox monastery that towered over Hapalmach Street just moments before the convoy approached.

As for the rest of what happened at about five thirty on the evening of September 17, 1948, in the streets of Katamon, a Jerusalem neighborhood that had been the ancient home of the holy man Simeon mentioned in the gospel of Luke, memories were perplexingly dim. Some of the people on the street explained that they'd turned away, unfortunately, at the precise moment. Others said that as soon as they heard the first shot, they'd run for cover; can you blame us? As for the Israeli police, they would've liked to have been of assistance, they'd offered with weak conviction, but if no one saw anything, you can't expect miracles. And so all the frustrated UN authorities could do was to attempt to piece together what had happened from the stunned survivors who'd been riding in the motorcade.

General Åge Lundström's chaotic account went like this: "We were held up by a Jewish Army type jeep placed in a roadblock and filled with men in Jewish Army uniforms. At the same moment, I saw an armed man coming from his jeep. . . . He put a Tommy gun through the open window . . . and fired point blank at Count Bernadotte. I also heard shots fired from other points, and there was considerable confusion. . . . When we arrived at the hospital . . . it was too late."

None of the assassins were identified, and the case was closed without anyone being charged. As for the foreign press reports that the Stern Gang, a militant Jewish organization, had ordered the assassination because they viewed Count Bernadotte as a lackey of the British and the Arabs, the Israeli authorities shrugged them off. "Jerusalem is always full of rumors," said an exasperated spokesman for the government.

Paisley had seen it all happen. He had been riding in the sedan at the tail end of the motorcade. And it was in the aftermath—that night? the next day or so? the details are sparse—that Angleton made his pitch.

It was a shaky time, when the young man had no idea what he'd do next; for all he knew, his career with the United Nations had ended with the count's death, and so had his future. The dismal prospect of having to slink back to Phoenix, of being on his own again, of having nothing larger in which to invest his faith, would have been devastating. At sea, he'd be looking for something to which he could be moored.

With little discussion, Paisley signed on as a paid asset for the CIA. Which, Pete thought, was testimony to his old colleague's delicate persuasion and impeccable timing.

But what if, Pete considered in the next unnerving moments, it was Moscow Center and Paisley who had set the trap? And Angleton had walked into it with his eyes wide shut and signed on a double?

IN THE DAYS THAT FOLLOWED, Pete tried to put all his disquieting thoughts aside. He needed some detachment if he were to make his way through the muddle. And so, he'd later confided to a friend, on a bright morning he'd traipsed off with a merry swagger to wander through the primeval thicket that stretched beyond the Bois de la Cambre. He had his binoculars raised and was diligently making a mental inventory of all the birds he spotted.

The forest teemed with life. A lively chickadee was chirping from the crook of a long branch. Its blue feathers and bright yellow underside made it easy to spot. On an adjacent tree, a magpie with its unmistakable fat white belly perched in seeming contentment. For Pete, the morning passed blissfully, full of a very satisfying enterprise.

Only that night, Pete would later share, he was unable to sleep. He lay awake in bed pondering his dilemma. How, he asked himself, do you identify a traitor? What kind of markings do moles have? What are their unmistakable signs? *What,* he wondered as he stared out into the darkness, *am I missing?*

Anatoly Golitsyn, the KGB officer whose defection in 1961 revealed a world of far-reaching intrigues.
Government Open Source

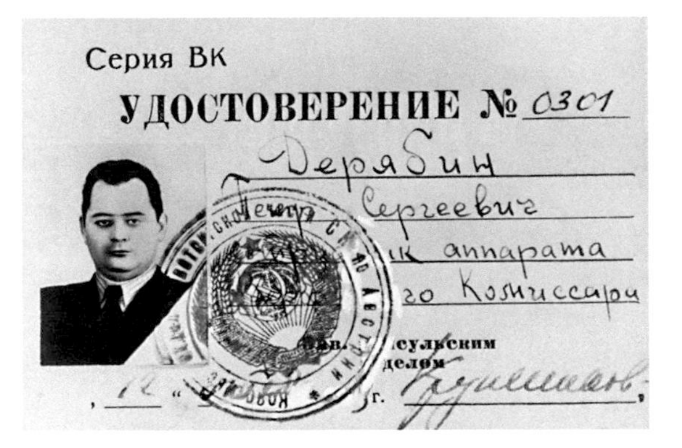
Серия ВК
УДОСТОВЕРЕНИЕ № 0301
Дерябин
Петр Сергеевич
аппарата
Комиссари
делом

The ID card of Pyotr Deriabin, the KGB major whom Pete Bagley smuggled out of Vienna in a machinery crate aboard the "Mozart Express" (*below*). Later known as "Peter," and working as a CIA analyst, he played Cupid for two young intelligence officers.
Wikimedia Commons

Yuri Nosenko, the KGB officer whom Pete Bagley first met in a Geneva safe house in 1962 and who became the enigma energizing his quest, his "Rosetta stone" that demanded to be deciphered.
New York Times

Pete Bagley, the retired spy who headed off on a mission to unravel "knot by knot these twisted strands."
Bagley family photo

When Nosenko suddenly resurfaced, Bagley, in disguise, greeted him with a brush pass, covertly executed in the foyer of the ABC Cinema in Geneva.
Geneva Library

Marti Peterson, the first female case officer at Moscow Station, and the Soviet driver's license she acquired so that she could drive her boxy Zhiguli car to the dead drops left by the spy code-named Trigon.
Peterson family photo

James Angleton, the controversial spymaster who, Pete believed, was more sinned against than sinning.
Associated Press

Ray Rocca (*left*), the young OSS sharpshooter who hitched his wagon to Angleton's star, and who in time became the counterintelligence chieftain's right-hand man (*right*), his hunter of "spirals."
CIA Open Source

Oleg Penkovsky was run by the MI6 and the CIA, delivering a treasure trove of microfilmed secrets before his arrest in 1962 and subsequent trial and execution. But Pete wondered when the Soviets had first gotten onto him.
Wikimedia Commons

John Paisley, the high-flying CIA officer who stirred dozens of top secret pots, receiving a medal from CIA Director William Colby.
Government Open Source

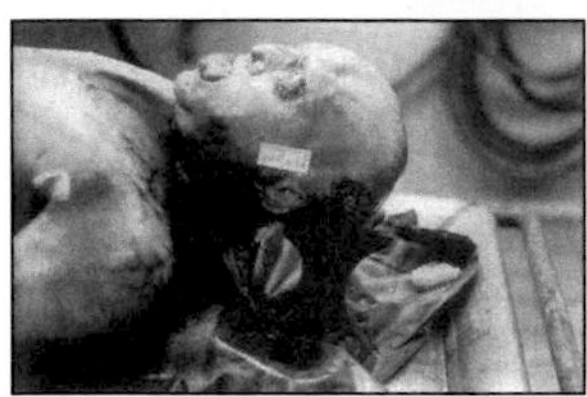

The decomposed corpse recovered from Chesapeake Bay that was officially identified as Paisley. But was it?
Government Open Source

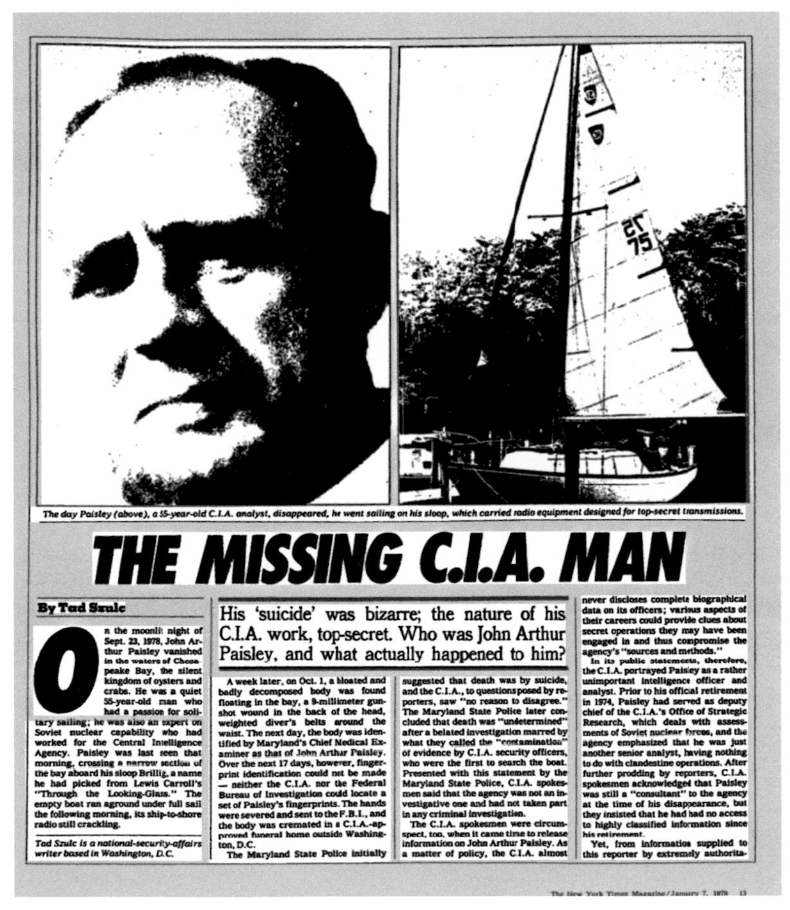

The day Paisley (above), a 55-year-old C.I.A. analyst, disappeared, he went sailing on his sloop, which carried radio equipment designed for top-secret transmissions.

THE MISSING C.I.A. MAN

By Tad Szulc

His 'suicide' was bizarre; the nature of his C.I.A. work, top-secret. Who was John Arthur Paisley, and what actually happened to him?

On the moonlit night of Sept. 23, 1978, John Arthur Paisley vanished in the waters of Chesapeake Bay, the silent kingdom of oysters and crabs. He was a quiet 55-year-old man who had a passion for solitary sailing; he was also an expert on Soviet nuclear capability who had worked for the Central Intelligence Agency. Paisley was last seen that morning, crossing a narrow section of the bay aboard his sloop Brillig, a name he had picked from Lewis Carroll's "Through the Looking-Glass." The empty boat ran aground under full sail the following morning, its ship-to-shore radio still crackling.

Tad Szulc is a national-security-affairs writer based in Washington, D.C.

A week later, on Oct. 1, a bloated and badly decomposed body was found floating in the bay, a 9-millimeter gunshot wound in the back of the head, weighted diver's belts around the waist. The next day, the body was identified by Maryland's Chief Medical Examiner as that of John Arthur Paisley. Over the next 17 days, however, fingerprint identification could not be made — neither the C.I.A. nor the Federal Bureau of Investigation could locate a set of Paisley's fingerprints. The hands were severed and sent to the F.B.I., and the body was cremated in a C.I.A.-approved funeral home outside Washington, D.C.

The Maryland State Police initially suggested that death was by suicide, and the C.I.A., to questions posed by reporters, saw "no reason to disagree." The Maryland State Police later concluded that death was "undetermined" after a belated investigation marred by what they called the "contamination" of evidence by C.I.A. security officers, who were the first to search the boat. Presented with this statement by the Maryland State Police, C.I.A. spokesmen said that the agency was not an investigative one and had not taken part in any criminal investigation.

The C.I.A. spokesmen were circumspect, too, when it came time to release information on John Arthur Paisley. As a matter of policy, the C.I.A. almost never discloses complete biographical data on its officers; various aspects of their careers could provide clues about secret operations they may have been engaged in and thus compromise the agency's "sources and methods."

In its public statements, therefore, the C.I.A. portrayed Paisley as a rather unimportant intelligence officer and analyst. Prior to his official retirement in 1974, Paisley had served as deputy chief of the C.I.A.'s Office of Strategic Research, which deals with assessments of Soviet nuclear forces, and the agency emphasized that he was just another senior analyst, having nothing to do with clandestine operations. After further prodding by reporters, C.I.A. spokesmen acknowledged that Paisley was still a "consultant" to the agency at the time of his disappearance, but they insisted that he had had no access to highly classified information since his retirement.

Yet, from information supplied to this reporter by extremely authorita-

The New York Times Magazine/January 7, 1979 13

The CIA, despite all their experience, were not very good liars, and that sent the press off on a dogged hunt for the truth about Paisley.
New York Times

Karl Koecher and his wife, Hana, the Czech secret agents who infiltrated the CIA while also having a "swinging" time living their cover.
Koecher family photo

The DIA didn't pay much attention to Nicholas Shadrin, a Soviet naval defector, while he was alive. But after his death, due partly to FBI and CIA blunders, he was belatedly honored.
Government Open Source

After the fall of the Berlin Wall and the toppling of Feliks Dzerzhinsky's statue in Lubyanka Square, Pete felt as if he were seeing "the other side of the moon."
Wikimedia Commons/ Associated Press

Sergey Kondrashev, the former Russian spymaster who helped Pete find the answers to the mysteries that had perplexed him for decades.
Wikimedia Commons/ C-SPAN

Chapter 25

YET WITH THE NEW DAY, Pete willed himself to forge on. *To bring old mysteries out of the dark.* That was the goal he'd set for himself, and, despite his vexation, he refused to surrender. Instead, he found reassurance in his conviction that he was no longer chasing a phantom; he now had a target. And with a renewed sense of commitment, Pete picked up the twisting trail of Paisley's life. Inevitably the prey will stumble and reveal himself, promised the old counterintelligence wisdom. The challenge, he reminded himself, was in hanging in there.

But no sooner had Pete returned to the hunt than he was confronted by a new perplexity: Paisley's incipient intelligence career had abruptly ground to a halt. He'd signed on with Angleton, but in the aftermath any of the usual telltale signs of covert activities had vanished.

Paisley didn't act as would be expected of an operative. He didn't remain in the Middle East. He didn't continue his UN employment. He didn't apply for a radio operator's job in government or the military. Instead, he fell in love, quickly married, and dropped out of any sort of life that would connect him to the secret world.

Suspicious, Pete doubled back and took a closer look at the wife. But he discovered nothing in her background of any operational significance. Maryann McLeavy was an attractive brunette who worked in the Manhattan office of a young swell who had presciently invested a portion of his significant inheritance to buy the Boston Red Sox baseball team. Unless she was spying for the New

York Yankees, Pete could see no path that would've brought her into the secret world. Adding to Pete's mystification, days after their wedding, the Paisleys moved to Chicago.

He enrolled in the international relations program at the University of Chicago, a student for a three-year master's degree that offered veterans academic credit for "life experiences" accumulated during the war. Yes, Pete acknowledged bitterly, working as a double agent, that'd count as a "life experience" all right. But this was only conjecture, Pete knew. And it was a supposition that was growing increasingly thin since, as far as Pete could tell, Paisley's life had turned pristine, to use the professional's adjective. There was no sign of his dabbling in any espionage work. Besides, what sort of opportunities would Paisley have now that he was at pasture in the groves of academe?

As his inquiry continued, Pete read that Paisley's wife had found a job in the university's chancellor's office. The chancellor grew so pleased with her work that the university offered her husband a full scholarship to ensure she'd stay on. Till then, Paisley had been scrambling to cover the tuition and the rent on their South Side apartment by driving a Checker cab nights and weekends. It seemed like a godsend, and Pete for a moment wondered about the true identity of Paisley's guardian angel.

But even as he mulled who was pulling the strings shaking Paisley's life, Pete saw that he had it all wrong. Paisley wouldn't accept the grant. In fact, he stubbornly refused to fill out the scholarship application. Indignant, Paisley growled that while he had grown up poor, he had never taken charity and he would not start now. He'd earn his tuition.

It was a logic that struck Pete as fatuous, but, he conceded, his own childhood had not been the struggle Paisley's had been. That sort of experience might very well have shaped a cockeyed pride.

So Pete plowed on. And just like that he found it. All at once everything fell into place.

Paisley couldn't accept the scholarship because he needed an excuse to continue working. He needed a cover story that would explain why every summer he left his wife and shipped out with the Merchant Marine. He signed on for voyages that took him to Eastern Europe and to Russia. And his foreign travels were apparently not restricted to school vacations, Pete discovered with a fresh exhilaration. "He used to take an incomplete in a course and take a job on a ship," a fellow student had told a journalist. "I think one time he missed an entire quarter."

Which made perfect sense, Pete considered. Especially if Paisley's handler had insisted he get a degree merely to add some sparkle to the cover he'd been fabricating. Of course the foreign voyages would have priority; they remained Paisley's primary occupation. They were the operational part of the secret life he was living. Everything else was building blocks for the legend.

When Paisley completed his coursework—a decade would pass before he finally submitted his thesis on the Soviet electronics industry and received his master's—his two disparate lives merged. He went off to Washington, spreading a fanciful story to friends about planning to knock on any door he could find as he searched for a job. He approached, however, only one employer, and the door was held wide open as he was welcomed in.

In December 1953, Paisley joined the CIA's Directorate of Intelligence as an economics intelligence officer.

This was the agency's "white side." It was the realm of the analysts, the agency's paper pushers and deep thinkers. It was territory that was very effectively cordoned off from the hush-hush spy work that went on behind closed doors in the "black side" Directorate of Operations. And it was the last place a mole would want to go to ground. Which left, Pete realized with sudden dismay, the case he'd been building against Paisley in complete tatters.

PETE'S THEORY HAD BEEN SHAPED by a single guiding hypothesis: the mole would be working hand in hand with the deskmen who ran field agents deep in Mother Russia. *He could only be in the Soviet Operations Division at the CIA*, he'd reasoned. And, his unflinching corollary deduction, the mole had to be someone near the top of the SB hierarchy. *Their source must have been in some stable position in that division*, he'd decided, *because most of its senior personnel were field operations officers likely at any time to be transferred abroad.* And Paisley didn't fit his once confident description at all.

At this low moment, Pete had a mounting sense that he was on a fool's journey. But doggedly he willed himself to continue. Yet at the same time Pete also couldn't help acknowledging that his perseverance was largely prompted by a single demoralizing reality: Paisley was his only candidate.

More distressing, the trajectory of Paisley's career on the "white side" sparked no reasons for renewed suspicions. He had started out in the CIA's Electronic Branch, and while he'd earned a reputation as an authority on Soviet electronic capability, this was not the sort of work that would have given Paisley access to agents operating behind enemy lines. Next, he was assigned to the National Security Agency at Fort Meade, part of the team assembled to make sense of the daily stream of information intercepted from a tunnel dug beneath the Soviet communication lines in Berlin. However, despite the ingenuity of the tunnel and the impressive size of the yield, the intelligence, it would be later discovered, was tainted. A Soviet mole in MI6 had revealed the plan before a shovel had gone into the ground and, therefore, Paisley's posting would have been of no operational value to the KGB.

After two years at the NSA, Paisley was brought back to the CIA, this time to the Office of Research and Reports. When Pete read this, he might very well have emitted an audible sigh; this was, his own memories made vivid, the dull roost of the agency's report-writing owls, a fiefdom as bland as its name. Langley's own Serbia, he recalled, a place for an officer to serve out his exile until it was time to retire. It would not offer access to the juicy details of ops behind enemy lines that Moscow Center so intently craved.

As Pete read on, however, his own misconceptions became apparent. He began to realize he'd been harboring a much too narrow view of Research and Reports. His initial prejudice had been borne out of a fieldman's swashbuckling pride that only those who lived with danger around every turn possessed the sort of intel that'd be valuable to the opposition. He began to see with increasing clarity that he'd been very wrong. For here was a "white side" officer who nevertheless was knee-deep in covert operations.

For starters, there was Paisley, using State Department cover, spending months at a time in Eastern Europe. His ostensible mission was to rumble around trade conferences trying to feed bogus stories about the capabilities of American technology to the Soviet representatives who showed up at these shindigs. But as Pete knew, the perfect double serves two masters, and the trick is to serve them both well. While Paisley was very efficiently spreading tales about the woeful limitations of America's electronic proficiency on orders from the CIA, he also had plenty of opportunities in his travels to give a more accurate assessment to a KGB handler.

His reputation enhanced by his success in these overseas missions, Paisley's agency career hit the fast track. He was assigned to do analytical work on Soviet strategic problems in the Office of Strategic Research. This, too, was a deskman's job, but it gave him access to the Crown Jewels—the sources and methods the agency employed to acquire intelligence on Soviet nuclear developments. It was a world of secrets—photos and electronic intel from spy satellites circling the globe, eyewitness reports from agents embedded in the Russian hierarchy.

As his responsibilities grew, Paisley, always ambitious, demanded even more access. And he got it. He was, as they say in the agency, soon cleared for the world.

Pete, fascinated, followed how Paisley, although still a deskman assigned to an office a world away from the Soviet Division, shrewdly managed to wangle his way into many of the ops the

Russia desk was running. In order to get a better understanding of what the Soviets were up to, Paisley insisted, he needed sit-downs with the defectors from Russia and the Eastern Bloc. And once again an accommodating agency agreed. He was welcomed into Camp Peary and Ashford Farm.

Paisley also succeeded in getting access to the HUMINT that agents were sending out of Russia. Pete, who knew from experience how things were done at the SB, was confident that the distributed reports would never have revealed the name of the agent. In fact, prior to distribution, the take would be rewritten with the intent of disguising all clues that might help identify the double working inside the enemy's house. Yet Pete expected that it wouldn't have taken much effort for a canny puzzle solver like Paisley to analyze the information the agent had reported and then figure out where he worked and what his responsibilities were. With the workplace narrowed down, the trail could eventually lead to a name. But if Paisley were still stymied, Pete had a good enough sense of how things were done in the secret services to believe a respected CIA analyst could've reached out to his old colleagues at the NSA with a convincing reason for why he needed to see the raw reports agents had transmitted. Still, even if all else had failed, once the HUMINT had been passed on to the KGB, they'd have no trouble in coming up with the traitor's identity.

A sudden chilling thought: Was this how Penkovsky and Popov had been identified? It would make more sense than the grotesque stories the KGB had been purposely spreading, the preposterous claims that their watchers had gotten lucky.

Pete now understood that he'd been wrong, at least in part. It wasn't necessary for a mole to be embedded in the SB Division to have knowledge of its secrets. Operating on the "white side," Paisley had managed to get all the access required to turn the agency inside out. And as for Pete's secondary hypothesis that the mole had to be someone who worked out of headquarters, Paisley, Pete noted with a small sense of gratification, fit that to a T.

Encouraged by these findings, Pete returned to the records. Only to be immediately blindsided.

THE YEAR WAS 1969, AND it was a contentious time in the Office of Strategic Research. The division's hard-nosed dispute was with no less a personage than Henry Kissinger, President Nixon's all-powerful national security advisor. At stake was a far-reaching arms limitation treaty being negotiated between the United States and Russia, SALT I. And Paisley was in the thick of all the uproar, arguing to anyone who would listen that Kissinger was "cooking the books."

Paisley had been given the complex job of assembling a National Intelligence [Security] Memorandum—a NISM, to those who dealt with such matters—that would inform Kissinger's arguments at the SALT negotiations in Helsinki. He went to work, and central to his findings was a discovery about Russia's hot-off-the-drawing-board SAM V missile. It had been deployed around major cities and, Paisley argued, it was a game changer: Russia was now armed with a very effective antiballistic missile (ABM). In strategic terms, that meant the Soviet Union could take America's best punch and still throw a hell of a wallop in retaliation.

Except Kissinger didn't see things that way. He dismissed the Soviets' ability to deter US missiles. He railed that Paisley was giving the Russian scientists too much credit; they just didn't possess the technology to pull off a sophisticated missile defense, and certainly not one that would leave sufficient resources to launch a retaliatory strike. America could pretty much wipe the SAM Vs out from the get-go. And, Kissinger further argued, Paisley's unfounded and alarmist mindset served only one purpose: it made an arms pact between the superpowers impossible.

Making his way through the rubble of this dispute many years later, Pete, while he had no trouble recognizing the weighty concerns that hung in the balance, still found himself perplexed. Why should Paisley advocate that Russia was a greater tactical danger to the United States than it actually was? It made no sense. Unless of course, he acknowledged, Paisley was precisely the CIA analyst he claimed to be and was responsibly going about his job as he saw it. But then why would Kissinger, who conceivably had no reason to diminish the danger Moscow's missiles presented, dismiss the ABMs as having more bark than bite? Kissinger and his team certainly would have known the facts.

Pete thought it all through, his ruminations reflecting the wilderness of mirrors that filled every counterintelligence agent's restless mind. And by the time he was done, he'd imagined a myriad of explanations. Paisley, he conjectured, could have been building cover, playing the hawk in a battle he knew Kissinger would ultimately win. His right-wing stance was a shrewd ploy to disguise his genuine sympathies; that, after all, was the card another mole, Kim Philby, ostensibly an MI6 hard-liner, had successfully played for years. Or, for all Pete knew, Moscow hadn't wanted the negotiations to succeed. But if the blame for the talks falling apart could be pinned on the United States, then the Russians would walk away with two victories. And maybe, Pete also had to wonder, it was simply a matter of Paisley's determination to do both his jobs as best he could. The CIA had assigned him to write a NISM, and, like the lover who is loyal in his fashion to both his spouse and his mistress, he had gone about his work.

In the end, Pete acknowledged it was one more mystery whose solution hovered beyond his grasp.

YET IN THE AFTERMATH OF the fiery dispute over Soviet missile capabilities, an even more perplexing event occurred: Paisley, according to several reports, threatened to quit the agency.

Why would he do that? Pete wondered. Had he had enough

of the demands of a double life? Could he no longer deal with the push-and-pull strains of answering to two opposing sovereigns? People who knew Paisley, Pete read, said he was a "nervous wreck."

But the CIA did not let him quit. Instead, he was sent off for a restorative year in London, enrolled for courses at the Imperial Defence College. It was the sort of posting, Pete understood, that the agency doled out to weary officers who needed a bit of time away from the frontline battles. Pete could almost hear the orders Paisley had received: Enjoy yourself. Get to know your family. You'll be back to the wars soon enough.

But as Pete read more about Paisley's activities in London, he began to suspect that Paisley had never left the front line. Clues jumped out from everything Pete encountered.

Paisley lived with his family in a CIA-rented flat off Grosvenor Square, a stroll from the US embassy. Yet rather than use the secure mail facilities at the embassy, Paisley had rented a post office box in Greenham Common; a gung-ho reporter had discovered this nugget in a CIA report written after Paisley's death that he'd retrieved under the Freedom of Information Act. The fact that this postbox was fifty miles from Paisley's flat set the alarm bells ringing in Pete's mind; he knew, after all, a good deal about tradecraft. And when he read that Greenham Common was home to a US nuclear base, the alarms started clanging as if they were signaling a five-alarm fire. Why did Paisley need a clandestine mailbox far from the watchful eyes in London? And why near a nuclear base? Was he communicating covertly with someone? And if so, whom? The out-of-town facilities had all the distinctive markings of an agent's dead drop. But what would Paisley have been covertly depositing? What might he have retrieved?

The unanswered questions spiraled down to another even more provocative query: On whose orders was Paisley acting? Was his control Angleton? Or a KGB officer in Moscow Center?

Pete didn't know. And all his speculations brought him no satisfaction. But he refused to abandon the hunt.

AND SO LATE AT NIGHT in Brussels, as Pete remained glued to his desk, he turned his attention, as he would tactfully put it, to "a deep and complex story." It was an agency operation that had crashed into shambles, leaving one man dead in the confusion. And Pete was drawn to the story not only by Paisley's active role in the affair but also by the bumbling participation of the very security officer who had been Nosenko's champion, as well as his own nemesis.

Chapter 26

1966

DURING THE EARLY DAYS OF the war, just after signing on with the OSS and years before he'd started working with Pete, Dick Helms had learned how to fight for his life on the green fairways of the Congressional Country Club in Bethesda, Maryland. The newly christened spy service had requisitioned the facility with its expansive red-roofed clubhouse perched like a citadel above acres of gently rolling land to serve as its training camp. Hand-to-hand combat was taught by Colonel William Fairbairn, and "Dangerous Dan," as the admiring trainees called him, was a legend. His torso and the palms of his hands were tattooed with the scars of knife fights from savage years with the Shanghai police and the British Special Operations Executive commandos. He knew a dozen different ways to kill a man, and that was just with his thumbs. "There's no fair play," he'd never tire of lecturing. "No rules except one: kill or be killed." For a prim and proper Philadelphian like Helms, genteelly educated at a Swiss boarding school and Williams College, this hardheaded philosophy was a revelation. It focused his mind, he'd say, for the dark and menacing war against the Nazis.

Two eventful decades later Helms found himself once again at the Congressional Country Club, still the uncompromising warrior, but now the peril was World Communism. He'd taken up residence (the club having been returned to its well-heeled members at the war's end) while he navigated his way through the torrents of two considerable personal challenges. One was a messy divorce; his soon-to-be ex-wife, in addition to being a sculptress of some renown, was a shaving cream heiress, and she had marshaled a battery of lawyers for the proceedings. But this exhaustive battle was nothing when compared to the infighting in the fickle corridors of the Senate as Helms, who had been nominated by President Lyndon Johnson to be CIA director, waited anxiously for his appointment to be fully ratified.

And so it happened that early on a bright spring morning in June 1966, as a besieged Helms sought refuge in the Saturday *Washington Post* and his second cup of coffee in the club dining room, a white-coated waiter approached to announce that he was wanted on the telephone. It was a call that would, as subsequent events unfolded, demonstrate that Dangerous Dan's kill-or-be-killed maxim applied with equal lethality to a silent war as to a declared one. And it was a call received by an old friend that would leave an infuriated Pete wondering, as he reviewed the entire disturbing incident years later, whether a mole had precipitated the abduction and murder of a double agent.

MY NAME IS IGOR PETROVICH Kochnov, the caller had begun in a heavily accented voice as soon as Helms spoke into the receiver on that June morning. I work in the Soviet embassy in Washington. A trade diplomat. However, that is not my real job. I am a major in the KGB, assigned to KR.

Helms listened with a guarded silence. But he knew KR stood for Kontra Razvedka, the counterintelligence unit of the KGB's First Chief Directorate.

I have information that would be of interest to the CIA, the voice, a deep baritone, pressed on.

It was at this point that Helms considered hanging up. The call might be a provocation; the Russians would find sport in embarrassing the man poised to take over the CIA. And, for that matter, so would a mischievous journalist. But what if it were the real thing? Helms, still a fieldman in his professional heart, was beginning to feel the excitement of the chase.

How'd you know where to find me? he finally asked. It was an attempt to buy time as he formulated a plan.

The caller explained that he'd first telephoned Helms's home in Georgetown. His wife had volunteered that her husband could be reached at the club. A beat, then the caller added provocatively, I was sorry to hear about Mrs. Helms's accident. I hope her leg has improved.

Julia had recently fallen off a ladder while she'd been hanging a painting, Helms knew; and he was impressed that the caller was aware, too. And he'd known his home phone number. Helms was not yet convinced, but he did want clarity.

Why did you reach out to me? he asked. The answer wouldn't necessarily matter, but it might provide a sense of whether he was indeed speaking to a KGB major. Helms needed more collateral before deciding to take things further. The prospect of his nomination being undone by a rash decision held him back, too.

The caller appeared to understand what was at stake, and he worked hard to demonstrate his sincerity. Talking rapidly, as if he feared that at any moment the call would be terminated, he launched into an elaborate account of a previous attempt to contact the agency. He'd reached out, he said, to a CIA officer in Pakistan named Gus. They'd talked, but the caller conceded that he'd grown apprehensive and had never shown up for the rendezvous.

Gus, Helms knew, was the work name used by Gardner Hathaway, an officer who was in fact based in Pakistan. And who was an old friend. If it was a coincidence, it was a fortuitous one. If it was a lure, it showed someone had done his homework to set a trap for a prospective CIA director. A case could be made for either proposition, and in his churning mind he did just that in the space of seconds.

At this point, however, the caller refused to allow the vacillation to continue. A meeting or nothing, he abruptly declared. Soon or never.

Call me back at this number in two hours, Helms decided. I will tell you where we can meet.

WHAT HAPPENED NEXT, PETE WOULD decide, was the mistake that set all the others in motion. It paved the way for the inevitable disaster.

Yet despite his anger, he could also understand the logic, however wrongheaded, that had guided the decision. A dark cloud of vulnerability hovered over the CIA in the 1960s. In the aftermath of the betrayals of Popov and Penkovsky, doubts and suspicions were the order of the day. And Nosenko's hostile interrogations had only added to the contentious finger-pointing.

Angleton and Rocca were convinced that a mole had caused all the damage, and compounding this sense of betrayal was the belief that the traitor was operating out of the SB Division. That was why, Pete knew, he'd been put under the investigative microscope, why his career had been tainted by preposterous allegations. Still, with the calming distance of time, he could find the professional objectivity to, if not condone, at least understand the motivations that had led to his being considered the bad apple. After all, he, too, had reached the same fundamental conclusion as his accusers: a mole was embedded in the ranks of the Russia House.

Helms's first call after speaking with the Russian was to Angleton. And with the clock already ticking—Kochnov would be telephoning Helms in less than two hours—Angleton worked, Pete acknowledged, with an impressive swiftness. For openers, he made sure the agency housekeepers had a safe house just a short stroll from the Russian embassy in Northwest Washington up and running, the vodka on ice and the tape recorders voice activated. Then he reached out to the FBI; they'd need to have an agent at the meet since this was shaping up, at least in part, to be a domestic counterintelligence operation. And finally he got on the phone with both Rocca and Desmond Fitzgerald, the Clandestine Services chief. The three of them quickly agreed that it would be imprudent to inform anyone in the SB about the meet. Their mutual assumption was that the Russian Division had been penetrated; if the details of the meet were shared with its officers, the opposition might very well be on to the traitor before he walked through the safe house door.

Angleton would years later personally apologize to Pete for keeping him, then the SB Counterintelligence head, out of the loop; Pete would find the grace not just to forgive his old colleague but also to understand the caution. Yet to this galling day Pete couldn't find it in him either to condone or excuse Angleton's next "strange and fateful" moves.

It wasn't just that Angleton had chosen to ignore, Pete thundered with wounded professional pride, "that only in the SB lay the experience and knowledge needed to assess and draw the maximum from a source at this level." Nor was it simply that he'd turned instead to the Security Office, a unit normally assigned to following up clues of hostile penetrations of the agency; its plodding snoops would have no idea how to run anything as nuanced, unpredictable, and decidedly human as an agent in place. But most infuriating was the specific security officer Angleton had handpicked to act as Kochnov's handler—Bruce Solie.

Solie, in Pete's stern judgment, "had only a shallow knowledge of the Soviet scene, knew little about the KGB, and possessed no experience in handling foreign agents." In the months ahead Solie would further demonstrate his "ignorance" to Pete by clearing Nosenko, and in the bitter aftermath Solie would emerge as Pete's personal nemesis. But that disaster was still brewing. For now, Pete's fury was sustained by the completely indefensible decision to assign Solie to run Kochnov.

With only minutes to spare, Angleton had gotten back to Helms to relay all the operational details he'd set in motion—including putting Solie in charge. Therefore, when Kochnov telephoned the club at precisely two p.m. and asked for Helms, the die, for all professional purposes, Pete believed, had already been fatefully cast.

AGENTS MAKE DEALS. THAT'S THE push-and-pull dialectic of the trade. They offer a prize, but it comes at a price. Money, a new life in the West, the chance to be a secret hero—whatever it is, they all want something in return. The proposition that Kochnov laid out later that afternoon in the Washington safe house was, even Pete could see, a tempting one.

Kochnov's pitch was made to Solie and Bert Turner, a burly veteran counterintelligence agent at the FBI. Cleverly, he didn't begin by making outrageous promises; that sort of hype would've turned his inherently skeptical audience against him from the start. But with more sharp tradecraft, he proceeded methodically to lay out the ascending steps in what would be a long-term mission. That way, by the time he'd reached the endgame, they were already there in their minds, waiting for him. They were panting for the prize.

His proposal, according to subsequent reports on the meeting, unfolded like this: I am only here on temporary assignment, Kochnov began. My time in America will end shortly.

At this point, the two officers seated across from him suspected what would come next: Kochnov would announce he wanted to defect. And they'd already received their instructions: that could be arranged—but only if the Russian had something to offer.

However, Kochnov surprised them. He wanted to go back to Russia—as a double. He'd work for the CIA as an agent in place in Moscow Center.

That got their attention: A spy in the enemy's camp would be an intelligence triumph.

Only the Russian wasn't done. He explained that his KGB career path would inevitably bring him back to Washington. A year, two years? He couldn't be certain when the assignment would occur. But he knew this much: When he returned to the *rezidentura*, it'd be as head of counterintelligence.

The operational prospect of running the man who'd have access to pretty much all the mischief the KGB was up to in America would have been intoxicating. But Kochnov had even more up his sleeve.

Upon his return from his tour as CI head in Washington, he charged ahead with a sudden momentum, he'd be bathed in glory for all his intelligence triumphs while in America. And he'd be rewarded with a seat at the big table in Moscow Center. How high he could climb in the hierarchy, well, that was anybody's guess. Kochnov wisely didn't embellish.

But by this giddy point, it's safe to say that Solie and Turner were guessing all right. They saw where this was going: Kochnov's a Moscow Center officer roaming the corridors of power, but all the time he's our agent. The keys to the enemy's kingdom were being dangled in front of them.

The two men fully understood how the game would be played. For Kochnov to succeed, he'd need to demonstrate to his bosses that he's a master spy. His promotions would be the rewards for a job well done. That meant that they'd have to give him goodies to take home. Oh, nothing too good. But not total garbage either. The product would need to be sufficiently enticing so that the headquarters crowd would be clapping him on the back and shouting Kochnov for commissar.

If that was what was required to get a double embedded in Moscow Center, Solie and Turner indicated that they were confident it could all be arranged. They'd be the shadow godfathers nurturing Kochnov's brilliant career.

Then, just as the staggering vision of what might lay ahead danced through Solie's and Turner's minds, the Russian abruptly switched gears. He revealed for the first time what he wanted in return. And he made it clear that it mattered very much to him. In fact, he'd played the entire scenario long to build up to this request.

His masters, he said, had dispatched him on his current deployment to America with a single goal. If he accomplished it, it would ensure all his future success. It would be the catalyst for a long and brilliant career.

His mission: He was to recruit a Soviet defector who'd been hiding for the past six years in America.

The defector's name was Nikolai Artamonov, now living as a naturalized American citizen under the name Nicholas Shadrin. At thirty, married with two children, Shadrin had been the youngest commanding officer of a destroyer in the history of the Soviet Navy, a rising star. But he'd fallen in love with a Polish medical student nine years his junior, and the two fled across the Baltic in a small launch and sought asylum in Sweden. The Swedes passed him on to the Americans, who welcomed him and even gave him an important post at the Office of Naval Intelligence. But that position had been short-lived, and as Shadrin's knowledge about the Soviet Navy grew dated, he was transferred to a pencil-pushing job in the Defense Intelligence Agency.

The KGB had poked around and discovered Shadrin was miserable, stewing with resentment over his fall from grace, angry that his once important life was now shrouded in drabness. That was why, Kochnov explained, his bosses believed Shadrin would be ripe for recruitment. And as further inducement, the KGB was ready to offer him a deal: after serving for a short time as a defector in place, passing on whatever he could grab from the cupboards of the Defense Intelligence Agency, Shadrin would be allowed to return home to Mother Russia.

Help me, Kochnov asked. Help me to recruit Shadrin. Even if

he only pretended to redefect, that would be sufficient, Kochnov nearly begged, to appease his bosses. Help me, and I will be able to help you beyond your wildest operational dreams.

So why?

That was the first question Angleton asked upon Solie's return to the agency with Kochnov's astonishing proposal that the CIA assist the Russians in getting their hooks into a naturalized American citizen. Why would the KGB eagerly want the return of a defector who could no longer divulge damaging secrets about Soviet naval strategy and procedures? Even the American navy, apparently, had grown bored with what he had to offer and had shuffled him off to the boondocks.

One possible answer was that Shadrin's defection had been a colossal embarrassment to the Soviets. That such a distinguished naval officer would throw his career away to live in the West must have stirred fears that others in the military would follow in his traitorous footsteps. But if he could be brought back to Moscow, paraded about as living, breathing testimony to the fool's gold that glittered in America, that would help ensure discipline in the ranks. His return would demonstrate that duty to Mother Russia was not just a heartfelt commitment but also a practical one: life is better here than over there.

Another possible reason was vengeance. If the KGB could lure Shadrin back to Moscow to execute him, it would be stunning proof of the wide reach of the Soviet intelligence services. It would be a warning to every future defector: you'll never be forgotten, and you'll never be safe.

In the end, as Angleton mulled the Soviets' motivations, he decided with pragmatic disdain that it really didn't matter what game the Russians were playing. Kochnov, he and Rocca were certain, was a provocateur. He'd been dispatched by the KGB to poke around the CIA and discover how the agency's mole hunt was progressing and to get an explanation for Nosenko's sudden disappearance (the defector's internment at Camp Peary was still secret). If that was the game the Russians were playing, Angleton, always up for intrigue, was determined to play it back at them. He'd feed Kochnov disinformation, and at the same time he'd be trying to get a sense of what the Russians knew, and, equally valuable, what they didn't know.

But the FBI and Solie—he served in the CIA's Office of Security and didn't have to answer to Angleton—had a completely different take on the matter. They believed that Kochnov was the real thing. Even better, he was offering them the opportunity of a lifetime: the chance to get an agent operating inside Moscow Center.

And all it would take would be to get a defector, now an American citizen, to act as if he's playing along with the Russians.

Do what you want, Angleton finally agreed imperiously. You want to persuade Shadrin to cooperate, be my guest. Angleton, however, offered one bit of advice to the novice agent runners. Whatever you do, don't let Shadrin leave the country. Don't give the KGB the opportunity to grab him.

And so the decision was made. First, Shadrin was approached, and, after some coaxing, he agreed to pretend to cooperate with Kochnov; perhaps he saw that playing the role of a double agent offered the possibility of more thrills than his humdrum DIA job. And then Kochnov was led to Shadrin to make his approach.

"What do we have to lose?" said one of the FBI agents involved in the scheme with a philosophical shrug.

THE OPERATION WAS CODE-NAMED KITTY Hawk, the name borrowed from the beach on the North Carolina outer banks where an FBI agent was heading for his vacation, and it ran for almost a decade.

As for Kochnov and his heady promises of the glorious career he'd embark on following Shadrin's recruitment, that never happened. Not long after the introduction was made, another KGB man took control of Shadrin, and Kochnov returned to Moscow, never to surface again. Or at least the CIA never heard from him.

With hindsight, Angleton's instinct that the Russian had been a dispatched agent seemed perceptive. And so did the decision, made by Solie and the FBI, not to tell Kochnov that the CIA would be manufacturing all the intelligence Shadrin would be delivering. The purloined memos, the top secret documents, even the stray bits of gossip—everything Shadrin passed on to the KGB had been creatively invented by the CIA. He provided US assessments of Soviet naval capabilities, analyses of Russian missile tests in the Black Sea, provocative hints about an American spy on the Russian missile testing range—and none of it was true. A team worked for years churning out the disinformation meant to lead the Russians in bewildering circles.

The CIA officer Solie put in charge of vetting these deliveries was someone who had first met Shadrin during the defector's initial stay at Ashford Farm. His name was John Paisley.

WHATEVER YOU DO, ANGLETON HAD warned, don't let Shadrin meet with his Soviet contacts overseas. It was a caution, Pete felt assured, any SB hand would've seconded. But since there were no officers from the impugned Russian Division involved in Kitty Hawk, the decision was left to Solie and Turner. And they couldn't resist the potential intelligence windfall that was being dangled in front of them in December 1975.

Shadrin was ordered by his KGB handler to go to Vienna, where he'd be instructed in the operation of the state-of-the-art radio he'd need to use in the future to make his reports. And while in the city, he'd also meet his new contact in the United States—an illegal who had been living undercover in America.

Solie heard this and told Shadrin to start packing his bags. A look at a new KGB radio *and* the identity of an illegal—now that would be a haul. And he told Shadrin not to worry; he'd have his back. He and his people would be lurking in the background in Vienna, watching Shadrin's every move.

A few days before Christmas, Shadrin, accompanied by his wife, Eva, checked into the stately Bristol (to this day Pete's mouth would water at the memory of the boiled beef served in the hotel's dining room), just across from Vienna's Opera House. He'd cryptically explained to his wife that he had some business in the city, and by now Eva had gotten used to not asking too many questions. Anyway, she was looking forward to the week of skiing in Zürs her husband had promised once his business in Vienna had concluded.

At six thirty on their third night in the city, Shadrin kissed Eva goodbye and said he was off to meet a Russian friend on the steps of the Votivkirche, the grand Gothic church whose twin spires rose above the center of town. "We'll have a late dinner," he promised.

Shadrin never returned. And his body was never found.

AND THE BETRAYAL?

Had the KGB been set on vengeance from the start? Had that been the impetus for Kochnov's mission? Were Kochnov's scenarios about operating as an agent in place hollow inducements?

Or had the Russians at some point realized that Shadrin's deliveries were too good to be true, and therefore they weren't? Had they begun to suspect he was playing them? Passing on bogus goods?

Or had Moscow recruited Shadrin in earnest, milked him for what they thought was a productive run, and then, when he recklessly ventured close to their hunting grounds, decided the opportunity was too enticing—and they pounced?

Or, as Pete was now coming around to suspect, had the entire operation been a time bomb ticking silently away, waiting to explode in the CIA's face, once Paisley had become involved? And Solie's foolhardy decision to send Shadrin off to Vienna had convinced Moscow Center to shorten the fuse.

In its dreadful end, the complex Kitty Hawk affair despite all its interlocking circles of deceit, Pete judged, was further proof of his guiding precept: *It takes a mole to catch a mole.* It was a rule that held as much for one side as it did for the other. Shadrin, he suspected, had been betrayed by Moscow's man in the agency. It was still a theory, but it was growing increasingly solid in his mind.

Kill or be killed. Yes, that's how it was in the war. That was the lesson the OSS instructors had pounded into the recruits. And that's how, Pete grieved, it was for Shadrin and the other victims of this conflict, too. With the stakes so high, Pete knew he needed to make sense of things.

Chapter 27

It was later that same week, a sunny day, and as Pete climbed the steep steps to the top of the Lion's Mound, as the hill that rose above the Waterloo Battlefield was known, just being outside under blue skies was helping him to feel better about things he'd confide to a friend. The site had been constructed more than a century ago, piles of brown earth extracted from all over the battlefield and then compressed into a hillock, and there were 226 steps to the summit. The climb was taxing, but Pete liked challenges. And it gave him time to try to work his way through his thoughts.

What bothered him, he explained at the time and then later in his writings, was this: When the Russians were running Shadrin—that is, before they'd grasped he was delivering counterfeit goods—they had assigned him a handler; first Kochnov, and then another KGB henchman. That was traditional Moscow Center tradecraft: a mole working in tandem with another undercover agent. It's their method, always a duo, always a two-man team. Russian intelligence services, he knew, had been doing things that way since the 1920s, when they had first sent a mole into Estonia. "These are," Pete understood with a hard-won insight, "risky operations." And the Soviets certainly knew that, too. They wouldn't embed a mole without, he predicted, "some way to keep track of his progress, give early warnings of his problems, and if possible help him out of his problems." There'd need to be a guardian angel hovering about.

But who was Paisley's contact? If Paisley was the mole, then whom was he working with?

It would stand to reason that it would have to be someone in the agency. Only a CIA officer would be able to provide the sort of valuable timely intelligence Paisley would require. Only someone working on the inside would be in a position to pick up any hints that Angleton, or maybe the Office of Security, were hot on the deceiver's trail and closing in. Only someone close at hand could throw a lifeline before the traitor went under.

But there was one problem with Paisley's having a liaison at the agency, whether it was a handler or just a contact person, to help him out if things threatened to all come tumbling down.

It was impossible.

The existence of a single mole inside the agency was unlikely enough. That there could be two traitors working in tandem seemed not just improbable, but totally impossible.

Yet it was the only explanation that made any operational sense. Unless, of course, everything Pete had been thinking, all his mounting suspicions, had been, to use his detractors' harsh words, "paranoid delusions." Unless his critics had been right. And for all these years he'd been wrong about everything. It was a painful thought, and for one rebellious moment, he suppressed a surge of resentment against the institution he was protecting.

When Pete reached the summit, though, the barrage of nagging questions and self-doubts subsided. He gazed out across the patchwork battlefield where so many men had died, and his thoughts grew centered, as they always did when he was up here, on the sacrifices made in the name of duty and honor. And on the proud military legacy he'd inherited, and his own hard-driving patriotism, on what he owed, and to whom. His service was of a different sort than his father's and brothers', but his sense of commitment, of the rightness of his cause, was no less firmly held. Tall and erect, he stared out into the distance, his thoughts wandering chaotically through a twisting maze of doctrines and beliefs, sentiments that had, in their enduring way, served to bring him to this juncture and his hardening vision of treason.

Lost in private recollections, he remained very still. But his demons would not rest for long. A sudden realization gripped him. Without warning, he'd later relate to a friend, he was pulled back into the tumult of his quest.

At once it was clear in his mind: *Of course. There had been someone in place in the agency.* Someone who could've worked hand in hand with Paisley. Someone who could've played the role of confidant, babysitter, and guardian angel. An agent whose own KGB case officer had praised him as a "super-spy." An intelligence officer whose secret work during his twenty-one years in America was so valued by Moscow Center that after the FBI finally caught up with him, the Russians hurried to cut a deal to bring him home.

Karl Koecher, as Pete had previously learned, had infiltrated the CIA, working in a hush-hush translation and analysis unit in the Soviet section and later as a contract employee. He had access to pretty much whatever the SB was up to—agent reports dispatched from behind the lines, transcripts of bugged conversations, defector debriefings.

It would make perfect operational sense for Koecher to be working hand in hand with Paisley. As a team they'd be, Pete judged with some grudging admiration for the enemy, an extraordinarily well-placed duo. Both men were, after all, based in America. Hell, they both had already tunneled their way inside the CIA.

Pete had another jolting thought: His initial impetus, the motivation for his starting his journey to revisit the past, had been two seemingly unrelated suicides—Ogorodnik's (the agent code-named Trigon) in Moscow and Paisley's on the Chesapeake Bay. But if the espionage operations had been intertwined, then perhaps so were the deaths. Had I, Pete wondered, intuitively suspected this all along? Had I known from the start where this would lead at its end?

But suspicions and instincts were one thing, and proof was something much more elusive. He just needed to find a definitive link between Koecher and Paisley, a nexus where their complicated lives merged. He needed to prove his thesis by finding a connection between the two moles.

It was with a newfound cheer that Pete made his way down from the Lion's Mound summit. His optimism had been restored, and so had his faith in the rightness of his quest. The 226-step descent, he'd remark, was accomplished as if in an instant, so focused was his mind on his fresh task.

THEY'D NEED A COVER STORY, a good one.

It was this fundamental deduction that guided Pete forward in the intense weeks that followed his transformative soliloquy on the high crest of the Lion's Mound. Koecher and Paisley would require a reason that'd convincingly explain why their paths might frequently cross. And it'd better hold up to scrutiny. Their encounters must not raise suspicions. At the same time, everything depended upon Paisley's being kept constantly aware, on his getting the first hints that the walls were closing in.

Yes, Pete decided, the connection between their two worlds would need to come off as happenstance. It would need to be a masquerade that worked without trying too hard. If Pete was on the right track, they were two moles intent on fooling an agency of professional deceivers. That'd take some doing, all right.

Pete's first thought was to focus on where their professional duties intersected. Looking at their careers this way, he quickly saw that both men dealt with Soviet defectors. Paisley's interactions, however, were flesh-and-blood exchanges with the Russians themselves; he'd show up to pepper them with questions at Ashford Farm or, as in Nosenko's case, Camp Peary. Koecher's job was much more arm's length. He'd sit in a cubicle, headphones on, listening to tapes of previously recorded conversations.

Still, their responsibilities might have had enough in common to propel them into each other's orbit. Paisley could claim that something had occurred to him about his debrief, say, of Shadrin, and he needed to hear the actual tape to ensure his memory was not playing tricks. That'd give him an excuse to go running to the AE/Screen office in Rosslyn, Virginia, where Koecher toiled. A seemingly chance meeting in the corridor or in the men's room would go unnoticed. But if Paisley needed to come up with regular excuses for visiting AE/Screen, that could start people thinking. Wouldn't be long before people—it was a building full of spooks, after all—would recall he was often seen with Koecher, and the last thing you'd want to do is give professionals the dots they could connect into a pattern. And if a crash meeting became a necessity, a spur-of-the-moment rendezvous in Rosslyn wouldn't work; there'd be too many obstacles straining credulity.

Another piece in the puzzle: When Koecher lost his job as a translator at AE/Screen, he swiftly landed a new post in the Office of Strategic Research (OSR). Where Paisley was the deputy director of the three hundred or so people on staff. It might very well have been, Pete suspected, that Paisley had played a backstage role in getting the mole his new job. However, their responsibilities at the CIA research unit were too different to have allowed them to have regular contact. An office friendship, Pete's instincts told him, would've attracted attention.

Pete kept trying, but in the end he grew convinced that the link between the two men would not be found in their professional lives. Their CIA duties occupied quite different operational spheres; meeting up in the course of their work would never seem truly natural. And naturalness was the essence of good cover; it's what made the lies believable.

So what was it, then? What would be a plausible rationale for their getting together from time to time for a friendly chat? As he pondered, a wartime X-2 tradecraft lesson his mentor William Hood had drummed in during Pete's early service in Vienna informed his thoughts. Footsteps attract less attention than sneaking on tiptoes. By that, the crafty Hood had meant: Don't want people to notice? Then don't skulk around. Do it in plain sight. Pete now refocused his thoughts accordingly.

Where could the two men meet? Nothing contrived. Nothing feverish. Nothing that'd appear out of the ordinary. Something out in the open. Pete was now hunting for something social, not at all professional. Even spies, he'd be the first to acknowledge, had lives outside the office. Yes, something personal and congenial would hold. Innocent-seeming run-ins. Innocent-seeming conversations, that'd work.

He made a list of possible scenarios in his mind, and one by one he tried them out. Did they share a psychiatrist? Were they members of the same church? Attend AA meetings? Had kids in the same school? Each of these would've worked, and yet when it came to finding the actual overlapping circumstances, he kept drawing blanks. It was as if Paisley and Koecher lived on different planets. All they seemed to have in common was their subterfuge. What have I overlooked? What am I not seeing? Pete found himself asking once again.

Then he thought he'd found it. Paisley's life outside the office was built around sailing. Was Koecher a sailor, too? Did they dock their boats at the same marinas? Sail the same waters? Shop at the same marine supply stores? As Pete reeled off these new possibilities, he was convinced he was on to something. This had to be it. He'd soon have the proof he needed. He'd have the evidence that explicitly linked the two men.

Only Pete couldn't find it. Still, though, he wouldn't surrender. It seemed such a logical cover—one he'd have suggested if he'd been running both men—that he refused to abandon the hunt. He was certain that if he kept at it, he'd uncover two lives tied together with a sailor's knot.

Ultimately, though, Pete had to admit that he'd been completely wrong. Koecher wasn't a mariner. He didn't have a boat. He didn't, apparently, like boats. For all Pete could discover, Koecher had never even been on a boat in his entire life.

So what, then, did Koecher do for a hobby? How did he fill his free time away from the office? When, of course, he wasn't trying to undermine the West.

Pete returned to the books and newspaper clips he'd gathered on Koecher and started to make his way through them once again. He'd stay at it for as long as it would take. He wouldn't give in to inertia. The link had to be there; he'd find it eventually.

Yet it didn't take Pete long at all.

KOECHER LIKED TO PARTY. THE high-spirited, free-loving antics of Koecher and his wife had been leeringly documented in the press. "The Swinger Spy," he'd been christened in one salacious tabloid account. An itinerary revealing the couple's desires, their rambles through private clubs and house parties in the Washington area, their evenings spent in crowded rooms filled with bouncing breasts and naked haunches and mattresses shoved end to end, had been assembled by diligent reporters. Had Paisley been part of that world, too? Did the two men share common appetites?

By the perverse logic of Hood's tradecraft, a naked cover would be an effective one; it'd seem as if there were literally nothing to hide. A line from one of Pete's favorite poets rose up in his mind: *My object in living is to unite/My avocation and my vocation.* Yes, this would be one way of intertwining the discordant trajectories of an unruly life. Still, sex clubs, what a place to look for a mole, he glumly thought.

Yet that was exactly what Pete did—from a distance of course. And, he asked himself, was it really so outlandish a proposition? So improbable? If Paisley was the mole, his loyalties were already careening all over the place. His inner world would be a mayhem crawling with risks, deceptions, and secrets. Would it be such a stretch to discover that his sexual encounters were also chaotic? The unconventional was a very repetitious way of life.

And there it was: Paisley, according to the muck many journalists (the initial groundbreaking research done by an industrious Joe Trento) had raked, was, like Koecher, a frequent and enthusiastic participant in Washington's adventurous sex scene. He'd show up for uninhibited frolics at private homes, at more elegant gatherings orchestrated by the Capital Couples and the Virginia In-Place, as well as at the down-and-dirty backrooms of certain redneck bars in Prince George's County, Maryland. There were also reports of his hosting sex parties on the *Brillig*; "ten people trying to make love on a thirty-foot sailboat can get pretty intimate," one female guest had discovered.

Paisley, always opportunistic, even went into the business of operating a sex club. In May 1972, with investments from several of his CIA colleagues, he'd bought a derelict lodge about an hour's drive south of the capital in Washington, Virginia. The original idea had been to transform the decaying property into the centerpiece of a trendy ski resort. But the cost was crippling, and so Paisley and another CIA colleague came up with an alternative vision. Without the knowledge of their other agency investors, they started hosting parties in the rustic Rush River Lodge. For a while, this resourceful scheme seemed to work; the cash flow guaranteed that the sizable monthly mortgage would be paid. But the Rush River Lodge was too run-down to attract a return crowd (even the showers couldn't be relied on), and it eventually went out of business.

Pete read all this with a sense of disbelief. Why would Paisley take such risks? He had a high-level security clearance. He worked at the White House with the national security advisor. He reported to Angleton. Why would he jeopardize his clearance, even his job, for these sorts of shenanigans?

One reason was obvious: The heart does what the heart wants. Perhaps Paisley enjoyed his rutting about too much to give it up. Whatever the risks, he'd decided, it was worth it.

But there was another operational explanation: It offered perfect cover. Paisley could confer with Koecher, and no one would give the conversation a second thought; there'd be other matters keeping them busy.

When Pete compared the reports of the two men's escapades, everything fell into alignment. The link was there: Both men, according to the eyewitnesses intrepid reporters had tracked down, had been seen huddled together during the quieter moments of the long, hectic nights at the Rush River Lodge and the Virginia In-Place. Undoubtedly, there would've been opportunities for whispered exchanges amid the sweaty couplings at other gatherings, too. What would be more natural than two studs taking a breather, chewing the fat as they regained the vigor for the next round?

The tradecraft, Pete had to marvel, was quite ingenious. If his supposition was correct, if Koecher and Paisley were two moles working in tandem, then they'd certainly found a unique way to ply their villainous trade. A transparent bit of mischief that concealed something deeper. It was the perfect ploy.

THEN, JUST AS PETE'S INVESTIGATION was gathering steam, just as he turned to the next page in Paisley's wide-ranging career, he was left stunned. There's Paisley, now deputy director of the Office of Strategic Research, globe-trotting to high-level nuclear armament negotiations, checking out spy satellite photographs that stream daily into the National Reconnaissance Office, and wandering through the corridors of the Nixon White House. It's a time when, after the 1973 Arab-Israeli War, Paisley's sent to Israel to help set up the classified satellite communications system the UN peacekeepers would use; and on his return, he's the senior CIA man writing the Middle East briefing papers for President Nixon. It's a time when his intelligence value to the opposition would be at its peak. And what does the mole do? In 1974, he retires. At an energetic fifty-one.

It made no sense. In fact, it turned everything Pete had been thinking upside down.

Chapter 28

1972

EVEN WITH ITS ORNATE BAROQUE crown, Kutafiya Tower was one of the shortest of the ancient towers protecting the Kremlin Wall. Still, it was about the height of a three-story building, and beneath its brick fortifications were a gloomy warren of rooms that in czarist days had been used as stables. During Stalin's reign, however, the horse stalls were demolished and the space was converted into what the Russian intelligence service, with its predilection for the arcane, called the Special Purpose Garage.

The garage's singular (if not very special) purpose was to house the fleet of shiny limousines used by the Russian president and the members of the Politburo of the Communist Party. In the 1970s—the Brezhnev era that Pete was delving into—the state sedans were Russian-made ZiL-114s. These were formidable behemoths, weighing in at close to seven thousand pounds, and ugly, too; their squat silhouette had been inspired by the tanks that had defended Stalingrad, with a grille blatantly lifted from the American Chrysler Imperial as a small, albeit incongruous, concession to style.

But the 114s did have some pep. There was a V-8 engine with a two-speed automatic transmission that on the Ring Road circling Moscow could hit 110 miles per hour. In addition, it was also outfitted with flashy Western bits and bobs: power windows, power door locks, a remote-control driver's mirror, and even radio telephones so the Soviet decision makers could keep in touch as they were chauffeured about. And to keep the fleet of ZiLs purring, a crack team of mechanics, each diligently vetted by the KGB, was assigned to the Special Purpose Garage.

One of the mechanics was a CIA spy.

During four dangerous years, this agent would be the very HUMINT side of an inspired ELINT (electronic intelligence) penetration of Russia's ruling posse that would have significant ramifications. To begin with, it would trigger a sense of calamity among the hard-liners in the US espionage community and the White House palace guard. And in another consequence—and this was precisely why Pete was sitting at his desk in Brussels and digging into its history with relish as well as a confirming satisfaction—it would bring John Paisley covertly back from his very short-lived and very staged retirement. Summoned by his old CIA bosses, he became an insider in a highly classified government assessment of the apocalyptic thrust of American might, a mind-rattling inquiry into who'd come out the winner if the two superpowers started madly hurling nuclear weapons at each other.

Which, Pete knew even before exploring any further, was the last place you'd want to find a mole. And the first where he'd want to be.

In the annals of intelligence, the missions that actually accomplished what they'd been designed to achieve were in short supply. Between the idea and the operational reality, Pete had observed during the course of his career, more often than not fell a disruptive shadow. But Gamma Guppy, as the CIA plan set in motion in the early 1970s to eavesdrop on Soviet bigwigs had been code-named, was that rare op that ultimately delivered even more than its heady promise.

Gamma Guppy's technical elements, engineered by the scientific owls in Division D of the agency, were both clever and complex. Yet at its roots, there were two fundamental components. One was that the CIA asset embedded in the Special Purpose Garage made certain each day that the radio telephones in the limousines were set to an identical precise frequency. And, the second part of the scheme, an array of calibrated parabolic antennas perched on the roof of the American embassy on Novinskiy Boulevard swooped up the radio telephone signals from the limousines shuttling about the city and simultaneously converted them into easily decipherable transmissions. After that, translators working in the embassy (supplemented by the AE/Screen denizens based across the world in Virginia) would churn out English transcripts of the eavesdropped conversations.

At first the take was mostly gossip. Brezhnev moaning to his wife about his expanding waistline. A tipsy proclamation of love by a Politburo member to his mistress. Another official checking in on his ailing mother with obvious heartfelt concern. On its face, it was inconsequential trivia, but the agency analysts and profilers lapped it up. In the trade, you never know what operational alchemy will conspire to transform lead into pure gold; the rule was to throw everything into the file and then hope it will someday come back to haunt the opposition.

But in due course, the listeners got a genuine earful.

It was May 26, 1972, and the historic winds of superpower diplomacy were swirling about Moscow. After more than two years of bickering and horse trading, US president Richard Nixon and Soviet general secretary Leonid Brezhnev, two previously unappeasable cold warriors, sat adjacent to each other as they signed a game-changing arms limitation treaty. SALT I, as the agreement would be known, put an admonitory halt to the feverish escalation of nuclear weapons; it froze the already bursting arsenals of intercontinental and submarine-launched attack missiles, and it also limited the number of defensive antiballistic missile sites. And at that rosy moment in the Grand Kremlin Palace, as the honor guards' trumpets roared in celebration and effusive statements pledging fraternity and cooperation poured forth, network commentators earnestly assured the vast American audience watching the ceremony on live television that the world had become a safer place.

But earlier that very day the Gamma Guppy team overheard something that left many of them wondering. And, if they were being entirely truthful, very scared.

BREZHNEV FEARED HE'D MADE A grave error. The Soviet general secretary had left a final negotiating session with Kissinger and was now sitting uneasily in the back of his ZiL limousine trying to calm his pounding heart. Had it been a mistake to limit the number of existing Soviet missile silos? he worried. He'd made the concession believing it wouldn't count for much since he had a trick up his sleeve or, to be more precise, concealed in the missile silos.

His ploy: The deal he had hammered out with Kissinger had limited the expansion of existing intercontinental ballistic missile silos to 15 percent. The tacit assumption by the American negotiators was that the silos would continue to be stocked with the Russians' workhouse nuclear weapon, the SS-11. This missile carried a single warhead that packed a nasty punch; a direct hit could blow a city the size of New York to kingdom come. But the United States was willing to accept the doomsday risk since it believed an incoming SS-11 would likely be destroyed by a swarm of US antiballistic missiles before it detonated.

Brezhnev and his advisers, however, had come up with a way to tilt the odds in Russia's favor. The SS-11s wouldn't be in the silos. Instead, the launch systems would be loaded with the new SS-19s. These missiles had a similar stiletto-like silhouette as the older weapon, but there was one very significant difference in its payload: it had four multiple reentry independent vehicle (MIRV) nuclear warheads. This design meant that each SS-19 launched had four times the chance of hitting its target with a 600 kiloton (the equivalent of about 800 million tons of TNT) nuclear warhead. A horde of US ABMs might succeed in taking out one or even two of the warheads, but there'd still be others, untouched and zeroing in. And after all, just one would suffice to get the horrific job done.

But Brezhnev was now worried. He'd suddenly realized he didn't know for sure whether the SS-19 would fit in the existing silos. And what good was the bullet if it couldn't be loaded into the gun? If the SS-19 couldn't be covertly lowered into the silos, then he'd negotiated a very bad deal.

Close to panic—or at least that was what the listeners heard in his voice—he placed a call to Defense Minister Andrei Grechko and implored, "Can we fit the new missile in?"

Grechko, with a politician's instinctive caution, said he wasn't sure. He'd need to check.

Finally the defense minister called back. There was no problem with "the main missile," he stated confidently.

"Thank God," erupted Brezhnev.

The Gamma Guppy listeners, however, were stunned. America had fallen for some very high stakes Bolshevik trickery. If World War III broke out, the United States would be left holding the short end of the stick—assuming, that is, there'd be any sticks left to hold after the planet had been immolated.

WHAT COULD NOW BE DONE? That was the question buzzing around the hush-hush precincts of Washington once the gist of the alarming Gamma Guppy transcript had been discreetly circulated. The treaty, after all, had been signed. Staring into the face of an absolute, diplomatic protest would be futile, as well as embarrassing, and retribution would be a dangerous provocation. Still, the enraged policy makers agreed that this sort of fiasco could not be allowed to happen again.

Among the like-minded professionals who spent their days thinking about the unthinkable, a revolutionary project began to be championed. It was a program designed to ensure that the next time American negotiators sat down at the big table across from the Russians, they'd know all the cards their opponents were holding, including the aces they'd tucked up their sleeves. And the best way to accomplish this, suggested a well-connected cadre of die-hard conservatives serving on the President's Foreign Intelligence Board, would be through a competition.

With a confident nod to the time-proven capitalistic wisdom of Adam Smith, they argued that the formulation of American nuclear policy should be a contest between two opposing groups of wise men. Each side would make their case, and the most persuasive argument would triumph. And the prize? The winner would get to set the battle plan for World War III.

In May 1976, this factious proposal was presented to the new CIA director, George Bush; he had the final say on intelligence assessments of the Russian Bear. "Let her fly! Ok," he breezily signed off. And so the game began.

On one side—Team A, it was unimaginatively called—were the traditional analysts, the CIA officers who, year after year, had churned out the National Intelligence Estimates that deciphered what the Soviet military was up to.

The opposition—dubbed with similar creativity Team B—was an ad hoc group of government and academic pundits who, although sporting eclectic institutional affiliations, were tightly bound by a common trait—a pugnacious conservatism.

Beginning in August 1976, and continuing on through the new year, it was a no-holds-barred competition. "Absolutely bloody," gloated one Team B participant. "Sometimes we left them speechless."

The CIA analysts, by and large, were convinced America was winning the Cold War. The Russians, they argued, were so far behind in the grim technology of destruction that they'd never catch up. They just didn't have the scientific know-how or, for that matter, the rubles to pay the bill. Therefore, Team A argued, the time had come to cut deals like SALT I; it made good strategic sense to institutionalize the way things stood since the United States firmly had the upper hand.

This optimistic take on things had Team B fuming. CIA incompetence and weak-willed liberal politics had "systematically underestimated" Soviet might and "projected a sense of complacency unsupported by the facts." The reality, they shrieked, was that "the Soviet Union is preparing for a Third World War as if it is unavoidable." To explain the enemy's determinedly apocalyptic mindset, a Team B leader offered this doomsday logic: "The USSR could absorb the loss of 30 million of its people and be no worse off, in terms of human casualties, than it had been at the conclusion of World War II." The resilient Russians, he warned with a chilling fatuousness, would shrug off the human cost of the next war—a mere 30 million souls—as the price that had to be paid for world domination.

In the end, when all the tooth-and-nail battles between the A and B teams had been fought, the conservatives won by a decision—a strongly politicized one. Intelligence, whose value to government decision makers traditionally had rested in its objectivity, in the unadorned presentation of the facts and nothing but the facts, would now be infused with the hard-liners' sharp, doctrinal prejudices. Not only were the CIA's National Intelligence Estimates on Soviet forces revised to reflect a starker, more worrisome verdict, but Ronald Reagan's election in 1980 brought many of the Team B players into official power. And they quickly flexed their newfound muscle by trying (albeit with only limited success) to dismantle "an elitist" CIA.

Yet as Pete made his way through the history of the Team A/ Team B muddle, it was Paisley's prominent role in this internecine battle that had his attention. Adding one more cover story to a long list, Paisley ostensibly had retired to a humdrum job in a downtown accounting company. But he was never there; his phone rarely rang, and his secretary wasn't sure what he looked like. Instead, he continued to report to CIA headquarters. And he had a new position, one—and this was the only accurate way to measure power at the agency—where his opportunities to get his hands on top secret files had dramatically increased. Paisley, according to one of the CIA officers with whom he now worked, had access to "anything that he wanted."

His job? Paisley was Team B's CIA coordinator. It was his job to get Team B whatever—*whatever!*—documents and files they decided to review.

And this was what left Pete baffled. What was Paisley, a man whose suspected treason was growing more and more set in Pete's mind, doing working in concert with an unrepentant band of right-wingers?

THE ANSWER, PETE ULTIMATELY CAME to believe, was not a simple one. Rather, it involved several explanations that wrapped around one another like the strands in a double helix. As for the fact that the elements were often contradictory, that was to be expected; moles, even more than most individuals, contain multitudes. So, Pete theorized—

It was all about cover. And what would be better cover for a Russian spy than going to ground in a band of fire-breathing conservatives?

Or, why not conspire to get America all revved up about the Soviet threat? It was in the Kremlin's interest for a panicked United States to spend promiscuously, to devote an even greater portion of the gross national product to the military. The more that the enemy spent on guns, the less it could spend on butter. In the long run, the skewed priorities would weaken the opposition on its home front and work to the Kremlin's advantage.

Then again, it was no less likely that Paisley, living his complicated life, was once more dutifully serving two sovereigns, and doing his best for both of them. He could be the obliging liaison to Team B, getting them whatever they craved. No task would be too onerous. He'd industriously scavenge through classified files and documents to discover the support for their arguments. And at the same time that he channeled the secrets to Team B, he'd also be sending them off to Moscow Center.

But, as Pete kept reading, he discovered one more twist to Paisley's scheming. The *New York Times* had run a page-one story by David Binder revealing not just the existence of the two teams but also the unsettling fact that the National Intelligence Estimate had, as a result of the competition, been rewritten to cast Russia as a decidedly more sinister threat. The disclosure had caused an uproar; indignant liberal officials took to the floor of the House and the Senate to condemn the pernicious infusion of politics into the CIA's assessment. And the source of the provocative story? In the aftermath of the mysterious death on the Chesapeake Bay, a resourceful journalist from the Delaware newspaper that had originally broken the Paisley story wide open reached out on a hunch to Binder. And the *Times* reporter decided that he was no longer bound by the traditional iron bands of confidentiality. Binder revealed the leaker's identity—John Paisley.

Which left Pete once more mulling the recurring nature of duplicity. Paisley had managed to undermine his conservative teammates by disclosing Team B's existence to the press. Yet at the same serpentine time, he also was likely betraying the CIA by shuttling its secrets to the KGB. All of which, Pete surmised, helped make Paisley the perfect double agent.

AND IN HIS PROFESSED RETIREMENT, Paisley continued to be busier than ever. When Team B was disbanded, he moved on to still another cover job. This position was at the MITRE Corp., one of the many Beltway tech concerns that each year received millions from the classified Black Budget that funded intelligence activities. Yet at the same, Paisley continued to have access to CIA headquarters, possessing code-word clearances and a green VNE (Visitor

No Escort) pass that allowed him to wander about unaccompanied through a labyrinth of secrets.

Paisley's new assignment, too, was just the sort of task a mole would find invaluable. He was a contributor to the top secret manual on the KH-11, a spy satellite that had been designed to photograph the deployments of Soviet nuclear missiles. In January 1978, a former CIA officer would walk into the Russian embassy in Athens and sell a copy of this manual. It hadn't taken the FBI long to discover his treason; he was quickly tried and convicted.

But, and this was what spurred Pete's suspicions, starting back in 1975, the Soviets had been outsmarting our eyes in the sky. They knew precisely how to disguise and camouflage their missile silos. It didn't matter that the KH-11 could take pictures from a variety of angles or measure the shadows the missiles cast. The photo analysts would pore over the product—code-named Talent—and throw up their hands in frustration. The Russians had clearly figured out a way to render the high-flying eyes peering down at the sites as good as blind.

Which meant either the Russian scientists were a lot smarter than anyone in the CIA imagined or they were getting spoon-fed the sort of inside information that made them look like geniuses.

STILL, EVEN AS EVERYTHING SEEMED to be going swimmingly for Paisley in his furtive retirement, Pete couldn't help but notice that the rest of Paisley's world was in turmoil. The signs were all there, but they brought Pete no satisfaction. He read the reports of Paisley's behavior with an empathetic sense of dismay; after his long quest, the hunter could not help but feel sorry for the hunted.

The list of Paisley's recent indiscretions was long and varied, but they all boiled down to this: "John was coming unwound," recalled a shaken CIA associate of Paisley's. "His life was falling apart at the seams."

Chapter 29

And Paisley? Just what was going on in his life?

Pete, the longtime student of intrigue, recognized the signs. They were flashing for anyone to see. How could the agency have missed them? The incomprehension was criminal. But his irritation soon subsided, and in its place he found himself being pulled back to a time in Cold War Vienna when as a young spy he'd learned this lesson the hard way.

An icy drizzle had been falling, but from his observation post inside the Bristol Hotel Pete was comfortably dry and, more crucially, had a clear view of the car parked on the Ringstrasse. As for everything else about the operation that had unfolded during the winter of 1954, well, from the start, Pete had been wary. Vienna Station was attempting to recruit Boris Nalivaiko, a gregarious, hard-drinking KGB thug, and when the agent who'd been courting Nalivaiko called to set up a meet, the Russian had turned cagey. He didn't say no, but he refused to discuss particulars on the phone. Give my chauffeur a sealed note specifying the time and place, Nalivaiko countered.

The suggestion made no sense to Pete. For one thing, it's the rare Soviet official who'd have no qualms about entrusting his KGB driver with an envelope containing the details of a get-together with a Western intelligence officer. And, for another, he didn't like Vienna Station's being put into such a subservient role. Allowing the target to set the terms seemed unduly reckless, the sort of capitulation that could come back and bite.

Still, Pete pushed himself to look at things from the Russian's perspective. Maybe he was simply being cautious. Afraid of a trap. Who wouldn't want more reassurance before he took such a life-altering plunge? And Pete knew his place. He was a newcomer, a junior officer, and a very junior one at that. His bosses were veteran professionals; he'd spent rapt evenings listening to stories of their wartime exploits. In the end, he'd decided to keep his concerns to himself.

It was just as well. Despite Pete's misgivings, the note was passed to the chauffeur without a hitch. Breathing a sigh of relief, Pete walked out of the hotel.

And nearly bumped into the hulking Georgy Litovkin, Nalivaiko's loyal legman. Pete recognized him from the row of grisly surveillance photos fixed to the wall behind the Vienna Station chief's desk. Had Litovkin also been observing the exchange? Had the whole thing been a setup from the start? Or was there an innocuous explanation? After all, the Soviet headquarters was nearby. Was it really so odd to encounter a Russian on the street outside the Bristol, strolling by on his way to the office?

But Pete, as always, was suspicious of coincidences; the operational wisdom was they were rare in life, and even rarer in the trade. This time he alerted his bosses, only they weren't buying into his worries. Relax, they responded, more bemused than critical. Vienna's a pretty small town; there's no telling whom you'll run into. Can't start imagining dastardly plots every time you see a Russki on the street.

So a couple of days later, Pete was back working the next stage of the operation. This time he was doing rover's duty, at the wheel of a car cruising the streets surrounding the leafy borders of the Stadtpark. By the foot of the park's towering statue of Franz Schubert, a CIA officer would be finalizing the arrangements with Nalivaiko for a meet this evening. If all went according to the plan, the Russian would deliver a bundle of documents in exchange for a bundle of American dollars. Pete's assignment was to roam the neighboring streets and see if anything caught his eye.

And something sure did. Standing at the edge of the park was the increasingly familiar figure of Litovkin. The Russian was gazing purposefully in the direction of the Schubert statue.

Pete hurried back to the office, only to arrive in time to hear the recruiting agent boasting about how well things were shaping up. Nalivaiko had been weeping, tears running down his cheeks, the agent reported with amazement. He's ready to defect. "And he's telling the truth," the recruiter declared with certainty. "Or he's a consummate actor."

Oh, he's an actor, all right, Pete interrupted.

"What do you mean?" the station chief challenged. His tone was unusually gruff, but Pete chose to ignore the warning.

Instead, Pete excitedly shared what he'd seen. Why does a Russian preparing to defect have a babysitter hanging in the wings? he demanded. And then he waited for an answer.

But no one could be bothered to respond. Finally, the recruiter taunted, "So you've seen Litovkin again. Ha!" The message was clear: the young fieldman had better get a grip, or else start looking for a less nervy line of work. He wouldn't be the first agent to crack before he turned thirty.

That night, then, Pete was once more a watcher, his eyes fixed on the Gartenbau Café. The place was jumping, oodles of *Gemütlichkeit*, a packed dinner crowd. Black-coated waiters weaved in and out of the maze of tables while balancing immense trays piled with impossible loads of plates above their shoulders. From Pete's static post on the street, he had a perfect line of sight toward a cozy table for two near the window.

Tonight, too, red flags were waving frantically in his mind. Everything about this meet, he told himself, was wrong. He wouldn't have let the Russians pick the spot. And he'd never, *never* have agreed to a sit-down in a sector controlled by the Soviets. They had a motor pool building just a stone's throw away, and it was big enough to hide a platoon of armed troops.

But Pete hadn't played the naysayer; he didn't counsel caution. What'd have been the point? Who'd have listened? His bosses were already anticipating the commendations that'd be added to their personnel files. The prospect of landing a well-placed Russian defector trumped all his misgivings.

So Pete was an eyewitness as the op fell apart. The recruiter was leaning conspiratorially across the table in the crowded Gartenbau Café when without warning the Russian hurled a pitcher of beer in his face. On that signal, a circle of scowling Soviet soldiers armed with submachine guns fell into place around the two men. Other Russians deftly moved into position to block all the doorways. And at center stage, Nalivaiko had risen to his feet and was delivering a well-rehearsed speech overflowing with outrage. Flashbulbs started popping, and in the embarrassing days that followed the evidence of a blatant American provocation aimed at an innocent Soviet diplomat landed on the front pages of many European papers.

And from that night on Pete had learned his lesson. He vowed "never again." He'd never again, he pledged, surrender after only voicing meek protests when everything was screaming that something was very wrong. "It'd be a mistake," he'd lecture, "to shrug it off." When you get a whiff of "something unpleasant or undesirable," don't hold your nose. Follow the scent. It was this operational insight, in fact, that had sent him on his long, relentless quest for a mole.

After all his years running agents, Pete looked at the spectacle that was Paisley's messy life and knew, as any professional would know, that here was a Joe crying out for help. Here was an agent begging for someone to talk him off the ledge. The agency never acknowledged the warning signs. They continued to give Paisley more responsibility, greater access. Or maybe, Pete suddenly wondered, the honchos at headquarters simply found it more convenient to avert their eyes; that's what people often did when they stumbled onto an accident.

But Pete refused to look away. Instead, he stared straight into the anarchic disaster of a life unraveling. And the questions he now posed were, Whose life was becoming undone? Was it that of a loyal veteran officer finally succumbing to decades of living with the tense demands of his furtive profession? Or did it belong to a traitor who could no longer juggle all his conflicting impulses, a divided self flayed by the warring demands of his two opposing masters, as well as the differing versions of himself? Or was it that of the longtime double agent who suspected his jig would soon be up?

Pete could only hope he'd find the answer as he returned to the exhibits chronicling Paisley's descent. The cuttings in the manila folders on the wooden trestle table that served as his desk, he wanted to believe, would contain the evidence he needed to lead him to a solution. Yet he approached this very personal part of Paisley's saga with the reluctance of someone who knew he was setting out to eavesdrop on the last gasps of a dying man.

Chapter 30

IT WAS A NEW MORNING, and Pete was angry. His anger had been mounting for days, a slow-burning fuse that had been ignited shortly after he'd begun tracking Maryann Paisley and her role in the entire affair. He had started in rifling through the clips only to be blindsided by his discovery that Paisley's wife had worked for the agency. *In the Soviet Division.* Pete's old stomping grounds. And precisely where, his initial theory had postulated, the enemy would want to insert a mole. Her assignment was, as the insiders would've put it with a knowing nod, "extremely sensitive." And as if that wasn't enough, adding fuel to his ire and consternation were her ties to a man Pete, for all intents and purposes, had thrown out of his apartment in Brussels.

At a quick glance, Maryann Paisley's responsibilities might be dismissed as perfunctory, even tedious. A routine filing job in a windowless agency vault room, the only sort of work a short-term contract employee could hope for. But Pete, who knew firsthand how things worked in the SB, had a much better sense of what her duties entailed than the bland accounts carried in the newspapers. According to these straightforward reports, she had been assisting in the reorganization of the SB's entire agent rooster. Which meant, Pete immediately grasped with a shudder, that whoever worked clandestinely in the Soviet sphere, each and every CIA spy lurking in enemy territory, the entire covert garrison both now and in the past—Maryann Paisley had access to their names. And code names. As well as a miscellany of personal details—spouses, children, even their medical checkups. It'd be an inventory the enemy would've fantasized about getting their hands on.

In Pete's overheated mind, a new string of flashpoints fanned the flames: How did Maryann Paisley, a short-term contract employee, not a fully vetted CIA staffer, wind up with such weighty responsibilities? Had she been directed to apply to the SB Division? The agency had morphed into a behemoth the size of a Fortune 500 company, and yet when she goes for a job she lands in a room where, day in, day out, she's sorting through the holiest of holies. How did that happen? Who paved the way? Was it Paisley? Someone else at the agency? And at the bottom of all his imploding concerns was a treacherous thought: Was Paisley running his wife? Was Paisley his wife's controller? Were they working in tandem?

From his desk in Brussels, Pete could not even begin to tackle these notions. The trail had been buried years ago, and only someone who had access to agency files could dust off the footprints. But neither the years nor the distance prevented him picking up one of the press cuttings and learning the identity of Maryann's boss. And with that name, the spiderweb spread to entangle his own life. It led Pete back to the case that had started everything. The "Rosetta stone," he had called it.

HER BOSS WAS KATHERINE HART. Who also happened to be the wife of John Hart, the officer William Colby had appointed in 1976 to lead the team clearing up the shambles Bruce Solie had left after he'd looked into the Nosenko affair. With an earnest pledge of due diligence, Hart had come to Brussels to interview Pete. Even in retirement, Pete had remained the Cassandra warning anyone who'd listen that Nosenko was not who he claimed to be. And, furthermore, that releasing him from custody would be tantamount to setting an active Russian agent loose in America.

They had met in the same comfortable wood-paneled room that

was now Pete's command center for his mole hunt. But midway through their initial conversation, he'd realized Hart's promise was as hollow as his research. The verdict had clearly been decided before the trial. Hart had not even bothered to read Pete's lengthy, itemized report on Nosenko. Nor had he had any interest in interviewing the Counterintelligence Staff officers who had been poking holes into Nosenko's convoluted explanations for the past nine years. When Hart returned the next day for a new session, Pete's frustration broke loose inside him, and he wouldn't let the disingenuous visitor past his front door.

Pete, while disappointed and chagrined, was not surprised when Hart's report vindicated Nosenko, declaring the Russian "totally trustworthy." (Pete, however, thought that was an odd appraisal of an officer who you claimed had betrayed his own service.) But he'd been blindsided when Hart and his clique led a no-holds-barred campaign to discredit anyone who'd dared to impugn Nosenko. With a not very subtle twist of the institutional knife, Hart had called his study the "Monster Plot." Only the monster was not Nosenko, but rather those who were "not objective, dispassionate seekers of the truth." For example, Pete Bagley.

As a smear campaign, it was quite effective. Paranoia, Pete came to grasp, "was a charge always difficult to refute." His raging battle with Hart went public when Hart appeared before the House Assassinations Committee in 1978. The committee had asked the CIA to present an officer who could speak about what Nosenko knew about the murder of the president. There was, however, not a single mention of Oswald in his prepared remarks. Instead, for an hour and a half, Hart's unwavering testimony defended Nosenko and impugned the CIA officers who had challenged the Russian's honesty. And Hart aimed his most censorious fusillade directly at Pete, going after him by name. He claimed Pete had even planned to murder the defector. His assertions, Pete would fume, "attacked me viciously . . . though he knew it was all nonsense."

And now, years later, Pete for the first time saw the full worrying dimensions in the tangle of connections: he stared at an elaborate puzzle at whose center Nosenko was firmly lodged. There was Paisley, who had first been Nosenko's interrogator and later his buddy, sailing down to spend weeks on end in North Carolina at the Russian's new home. It was a friendship that struck Pete, who'd witnessed the temper in Paisley's interrogations at Camp Peary, as improbable as if he were to spend a weekend shooting the breeze with John Hart. Unless, of course, Paisley's original animus had been staged.

But there were more strands in this expansive web. Paisley's wife gets her CIA job from the wife of the man who'd been Nosenko's champion (as well as Pete's vigilant nemesis). And if that's not irregular enough, Maryann Paisley's assignment brings a lowly contract employee each day into the rarefied confines of the vault holding the Crown Jewels.

Pete was nothing if not stubborn, but try as he might, he still could not fit all these interrelated pieces into the single puzzle he was assembling in his mind. All he could say at this point—and he held this view with rising conviction—was that in this world of coincidences there had to be a trail pointing the way through a legacy of betrayals. The challenge was to find it.

AND IF MARYANN PAISLEY WAS entangled in this plot, then how was she dealing with a husband—and suspected double agent—whose life was whipping about in every direction? The short answer, Pete discovered, was that she was running for the hills.

Maryann, according to all reports, was not a fellow traveler in her husband's pursuit of communal sexual adventures. She was not a participant in the mischief that sent Paisley scampering to free-love parties throughout the Washington area. In fact, she was trying to disentangle herself from her twenty-seven-year marriage. In 1976, she had forced Paisley to move out of their spacious house on Van Fleet Drive in McLean, Virginia, and, her rekindled anger coinciding with her husband's specious retirement, was now talking with increasing firmness about turning their de facto separation into a formalized divorce.

So Paisley, almost a bachelor, retaliated by sowing his not very dormant wild oats by carrying on a public affair with Maryann's best friend, Betty Myers. They were pretty much living together, and they made no secret about it; Paisley's friends, even his kids knew. Myers, Pete read, rationalized this liaison by insisting that Maryann had "thrown us together." Maryann, however, looked at things differently. With the fury of a woman betrayed, she confronted Myers and, according to the deposition she gave in an insurance inquiry after her husband's death, said she never wanted to see or talk to her again. In fact, she now suspected that the romance had been going on behind her back for years; trickery, she was learning, was the single constant in her husband's helter-skelter world.

But her anger apparently didn't spill over to upend completely her teetering relationship with her husband. They had separate households, but they were still sleeping together. Paisley also spent nights with Myers, too. And duplicity was once again the recurrent pattern in this disparate life; neither the estranged wife nor the girlfriend was cognizant, as the professionals would put it, that he was bedding both of them.

Yet loyal to both in his fashion, Paisley went to the trouble of trying to demonstrate that each of the two relationships had a future. Although they were living apart, he attended marriage counseling sessions with Maryann. And with a similar fervor he dutifully went to couples counseling with Betty Myers. He wanted each woman to know that he was committed to her.

Which, Pete thought, was by itself proof that Paisley had all the makings of a double agent. And a good one, too, since he apparently kept them both pleased.

But while Pete was a man of the world and was not too judgmental about Paisley's escapades, he was also a doting father. And from what he found, he couldn't help but feel that Paisley's children had been forced to pay for their father's sins. Paisley's wayward impulses had knocked their lives off-balance.

It had been August 1973, a dull, humid summer night in the Virginia suburbs. So Eddie, Paisley's teenage son, and a few friends got into their cars and headed into Washington looking for fun. They had begun drinking on their way into the city, and at midnight when they were heading back home, they were apparently still at it; beer cans were littered about the cars.

At the wheel of his 1966 Ford Galaxie, Eddie led the way from Georgetown across the Key Bridge and onto the George Washington Parkway. Friends trailed in another car.

And as soon as the two cars were on the highway, they began jockeying for position, each trying to get the lead. The speed limit was fifty miles per hour, but they paid that no mind. All that mattered on a night filled with drinking was outracing each other.

Eddie pulled ahead. He looked to be the certain winner. But then, for reasons he never shared, he turned off onto Route 123. This was the exit that led to Langley, and the CIA complex.

But Eddie didn't get very far. As he headed down the twisting exit ramp, he lost control and smashed into a tree. Police reports would later estimate that the car had been going eighty, maybe even one hundred miles per hour at the moment of impact. His best friend, Brian Demmler, was killed.

Brian's estate sued the Paisley family, and the trial was an ordeal. Eddie, still devastated, made a terse confession to the judge: "We went downtown and we drank a little too much and came back racing down the G.W. Parkway. We were playing tag on the G.W. Parkway and I hit a tree."

The judge treated Eddie as a juvenile and he was paroled. But his father, by all the accounts Pete read, served a life sentence. "It was a terrible, haunting thing to him," said a close friend. And, Pete suspected, another blow that pushed Paisley's skidding life further off the rails.

Eddie's mother suffered, too. In her grief and mourning, she decided, her lawyer would later remark, that the time had come to get a divorce. She understood the crash had been an accident. Yet she also, she'd tell people years later, suspected that, just as shrinks are fond of saying, there are no accidents.

And she could no longer hide in self-delusions. The dislocation that characterized their lives needed to be confronted. The chaos at the core of their wayward family—the cloak of secrets that draped her husband's life—had made them all victims. She felt an intense and searing guilt for her tacit complicity in the wreck that was their fractious, hapless world.

Their daughter Diane, too, Pete read, had her own sort of unhappiness growing up in an unhappy family. Yet her father was apparently too preoccupied to come to the rescue. His shrugging attitude toward his children, an annoyed relative had explained, was, "I don't see them day-to-day, so when I do see them, why should I spend all my time disciplining them?"

Or for that matter, Pete railed, how could a man who couldn't apply any discipline to his own life straighten out someone else? Then again, a traitorous heart pledges no allegiances.

AND IN THIS WORLD BRIMMING with conundrums, there was also the enigma of Paisley's newly acquired bachelor pad: Why settle there? It was on the eighth floor a dull redbrick high-rise at 1500 Massachusetts Avenue in Northwest Washington. The commute from downtown to the suburbs of Langley, where he still had an office, was at rush hour a mind-numbing trek through bumper-to-bumper traffic. The rent, too, was no bargain, and the listings were filled with less expensive apartments in the neighborhoods adjacent to the agency. Further, undoubtedly a real blow to Paisley, the location rendered his ham radio gear pretty much inoperable; the lifelong enthusiast couldn't get permission to erect the type of elaborate rooftop antenna he'd need to pull in a signal.

But one possible explanation for Paisley's move did stick out, and it pointed Pete's mental compass in a now familiar direction. The apartment complex, it had been documented, was the nest for a swarm of "nightcrawlers."

As any officer who'd served in the SB knew, the nightcrawlers were specially trained KGB operatives posted under diplomatic cover to the downtown Russian embassy. But when the sun went down, they'd head out to prowl the fleshpots of Washington. They'd be on the lookout for government clerks, congressional aides, servicemen—anyone with a connection to official Washington. Once they had a target in their sights, the mission would be to lead the lonely stray into a honeypot. Hidden cameras would be clicking away to capture the intimate embraces. And in the morning, they'd offer the compromised victims a choice: either provide a continuing stream of secrets or gird yourself for ruin. Washington's gay bars were the nightcrawlers' favored hunting grounds.

In apartments just down the hallway from Paisley's new home on the eighth floor were eight KGB agents. Two teams of nightcrawlers, agents whose rough profession had them stealing secrets night after night, were lodged a circumspect stroll down the corridor from a CIA official who had unfettered access to a bounty of highly classified information.

But no less problematic to Pete was the KGB counterintelligence officer based at the embassy who ran the nightcrawlers. Vitaly Yurchenko, whose busy career would eventually earn him a general's rank in the KGB, was, according to FBI surveillance reports, a frequent visitor to the apartment complex. At the time, the bureau watchers figured he was keeping track of his blackmail teams. But looking at things now, Pete had a new suspicion. Was Yurchenko checking in on his mole? Was he the embassy officer running Paisley?

If that was the case—and Pete feared that it was—the professional in him found himself admiring the operational elegance of the enemy's subterfuge. Yurchenko could parade past the FBI cameras in plain sight. He'd be unconcerned. He'd probably wanted them to clock his appearances. The nightcrawlers were the linchpin of an elaborate deception scheme, a way to provide foolproof cover. For while the bureau would be gloating that they'd nailed a KGB security officer checking in on his stable of gorillas, they had no idea of his actual mission: Yurchenko had sauntered past their observation post on his way to a meeting with his mole.

But no operation lasts forever. No operative can keep on endlessly carrying the accumulated weight of his many lies. The time comes when the agent's nerves start to fray; he's jumping at sudden noises, seeing things in the dark. Or, just as likely, he confronts the irrefutable evidence that the thoughts keeping him up at night are the real thing: they're on to me. And scattered throughout the weeks leading up to Paisley's fatal sail on the Chesapeake Bay were a series of small whimpers that, to Pete's carefully tuned ears, echoed like a giant wail of despair.

So there is Paisley dropping everything to go off on a long winter's sail on the *Brillig*, only to return to port after two days. His heart wasn't in it, he told friends. Then he's contacting another old buddy and asking about the possibility of returning to the Merchant Marine; he needed to find a new line of work, he dramatically explained. Or there he is hectoring someone he'd bumped into at the suburban Virginia restaurant where his daughter worked. The guy was in the electronics business and Paisley's wondering if he had any contacts. I'm looking to change my life, Paisley implored. Maybe you know someone who could help me reinvent myself as the shipboard radio operator I'd been in my carefree youth.

And then there's Paisley's showing up without an appointment in the second-floor nerve center of the CI Division, cleared for the world Visitor No Escort badge dangling from his neck, and he's telling the head of the unit he's looking into leaks Capitol staffers had made to the KGB. Only without really trying the CI honcho later discovered that there was no authorized investigation into congressional leaks. Was Paisley, his nerves increasingly on edge, poking around to see if anyone was asking about him?

Or there was Paisley losing his wallet, and rather than calling the credit card companies to cancel his cards, he put a lost and found notice in the *Washington Post*'s classified-ad section. That struck some friends as a strange way to right the situation. But then a few days later, Paisley was spotted carrying his wallet and, wouldn't you know it, it was packed with credit cards. So perhaps he'd found it, or maybe it wasn't lost in the first place. Which would make sense if the ad's actual purpose had been to send a prearranged distress signal to his handler. Rescue me, he might very well have been pleading as his unraveling life flayed about. But wouldn't Paisley have used a workname if he'd been reaching out to the Russians? Or did he fear his hiding behind an alias would sound more alarms, bring confirming attention in case he was already being watched by an agency surveillance team? Once again, solid answers eluded Pete.

Yet all Paisley's varied moods and crisscrossing lives seemed to take a restorative pause on August 25, 1978. It was his fifty-fifth birthday, and Betty Myers had organized a party at Paisley's new apartment, inviting a handful of his friends. After the festivities, when she was alone with Paisley, a rare moment of what she'd call "lightness" seemed to animate a soul that lately, she'd perceived, had been troubled.

"Someone had brought balloons," she recalled. "After everyone left, he sat down on the floor and blew up all the balloons, tied them, took them up in his arms, and opened the window onto Massachusetts Avenue and pushed them out. And they flew all over the park and that was a wonderful things to see him do."

But this respite was short-lived. On the very day when Paisley had experienced a small sense of rightness and well-being, new storm clouds had started gathering. And Pete discovered to his surprise that all along he'd been walking in someone else's footsteps. For on Paisley's birthday, a CIA officer informed the agency's director of the Office of Security that he "believed John Arthur Paisley, the former Deputy Director of Strategic Research, was working for the KGB."

Chapter 31

1978

David Sullivan was trekking through the long, lonely desert of the disgruntled professional. His concerns had been ignored, and over time the slights had kept working away, breeding new suspicions. As an analyst assigned to the high stakes Team A vs. Team B competition, he'd been a hard-liner grumbling that the agency had cooked the books. Then he was the lone voice calling out in the intelligence wilderness charging that the only way the Soviets could've managed to conceal their nukes from our spy satellites was if they'd gotten their hands on the operational details of the flights. And no sooner was the SALT I treaty signed than he was letting everyone know he'd unearthed evidence proving the wily Russians were cheating with abandon. Yet in return for all his conscientious warnings, Sullivan would complain, "I had gotten nothing but resistance from the top brass."

Exasperated, he took a step that would be his downfall. Since no one in the agency was listening, Sullivan decided he'd make his case to a more sympathetic audience. He handed the report he'd written on the SALT violations to an aide of Senator Henry "Scoop" Jackson, a legislator who believed every treaty with the diabolical Russians was a fool's game. The CIA director learned about the unauthorized disclosure and, according to an official quoted in the *New York Times*, "hit the roof." It didn't matter to him that both the senator and his aide had high-level security clearances;

no bumptious analyst can take it upon himself to hand out agency reports to the Hill.

By the time Sullivan was summoned to the Office of Security to get his comeuppance, his temper, too, was blazing. He had had enough. And he decided the time had come to share what had been building up in his mind.

SULLIVAN WENT TO WAR, AND as Pete avidly turned the pages recounting an interview Sullivan had given to a reporter in 1988, he very well might've wished he'd been there to fire away, too. For Sullivan, on that morning in August 1978, looked Robert Gambino, the agency security chief, directly in the face and with no pretense at hemming and hawing shared Pete's long-gestating nightmare.

"I think there is something rotten in Denmark in this agency," he declared in a burst of anger. "Too much shit has gone wrong. There should not be this much resistance to my work. My work ought to be applauded."

Gambino did not respond.

Undeterred, Sullivan plowed on. "We have lost spy satellite capability, human intelligence, and signals intelligence. Do you think all of this is by accident?"

Gambino chose to treat the question as rhetorical.

"You have got moles here," Sullivan blurted out at last.

He gave Gambino a list of ten names. But his chief suspect, he stated, was John Arthur Paisley. "I am convinced he is the mole."

Like the director, Gambino had no truck with an analyst who made his own rules. He'd been prepared that morning to throw the book at Sullivan. And in due time, he would; Sullivan's days at the CIA were numbered.

But just as Gambino knew his job, he also knew his priorities. There was no responsible way he could ignore the charge that had been made; what he personally thought of Sullivan, the reckless journeyman with an inflated ego, was irrelevant. There were ten names that needed to be investigated at once. And one that would get special attention.

Once Gambino and his team went to work, they soon discovered that Paisley had not been polygraphed since 1953. In itself, this was more an embarrassment to the Office of Security than any indication of improper conduct on Paisley's part. The usual practice was for officers with Paisley's level of access to be hooked up to the machine every year, or certainly every two years if no one was on top of things. But this sort of lapse in the vetting process—*twenty-five years*—was most unusual. It needed to be rectified at once.

On Friday, September 22, 1978, Paisley received a letter from the Office of Security. It explained that individuals with high-risk security clearances were required to undergo a full background investigation on a regular basis and asked for Paisley's cooperation in the process. As the holder of a top secret security clearance, a Single Scope Background Investigation that covered the past ten years would be necessary. Spouses, neighbors, and work associates would be interviewed. Tax returns and banking records would be reviewed. To start the process, please fill out the enclosed personal history form. After it has been received and reviewed, you will be contacted to make an appointment for a polygraph examination.

The letter was the standard notification sent to agency officers whose clearances were undergoing review. Its tone was officious, but banal.

Yet apparently that was all it took.

Two days later, Paisley cast off the *Brillig* from its mooring on an isolated cove in Lusby, Maryland, at the broad mouth of the Patuxent River. He sailed into the calm, dark waters of a windless Chesapeake Bay. And he was never seen alive again.

WILL I EVER KNOW WHAT happened? Pete wondered. Will I ever succeed, he asked himself, "in unraveling, knot by knot, these twisted strands?" He had sifted through the evidence, applied all his substantial powers of reason and analysis, and in the end he still had more questions than answers about what had occurred on the *Brillig.*

Why? The classified agency documents in the sloop's cabin. There was no reason for the secret papers to have left headquarters, let alone be on a "retired" officer's sailboat.

Why? The burst transmitter in the cabin. The CIA stated it had not given Paisley this sophisticated piece of electronic equipment. Then who did? It was a device used to transmit or receive tens of thousands of words or signals on a preset frequency, a scrambler that sent coded communications to satellites. Secret agents used it to keep in covert touch with their services. But with whose intelligence service was Paisley communicating? Whose satellites were receiving the signals?

Why? The bogus *Washington Post* employee card found on board. The newspaper shared an alleyway with the Russian embassy. Did the pass allow Paisley to sneak through the *Post* building for meets with his handler?

Why? The live cartridges found scattered on the sloop's deck. From whose gun? Had there been a struggle? Where was the gun? There was no weapon found on board.

Why? The bloated, decomposed body found eight days later floating in the water just east of the mouth of the Patuxent River, an oozing bullet hole behind the left ear, two sets of heavy diving belts wrapped around the corpse. But would Paisley have committed suicide? For what reason? He had no financial problems. His wife and girlfriend said his moods went up and down, but he didn't appear to be deeply depressed; there were "no clues," they grieved. And could anyone have committed suicide in such a fashion? Wrapped in thirty-eight pounds of chains, and then shooting himself in midair as he jumped from the deck? And the bullet wound was behind his left ear. Would the right-handed Paisley have been able to contort himself in the midst of his jump to get off such an awkward shot? And why the diving belts? Was this part of a grand scheme to disguise the suicide? To ensure that the body would lie forever at the bottom of the bay? That it'd never be found? But that wasn't necessary; the payments on his two insurance policies were guaranteed even in the case of suicide. Besides, where did the second set of diving belts encircling the corpse come from? Paisley, insisted everyone interviewed, owned only a single belt. Whose were they?

Why? The body was cremated in great haste. In fact, it'd been burned to a powdery ash before any of the family had even identified the corpse. And the autopsy was perfunctory, as well as problematic. Both the CIA and FBI notified the Maryland state police that they—astonishingly—had no fingerprint records for this veteran agency officer. And a visual identification had been impossible; the body had been grossly distorted by its week in the water. Another problem: the body hauled out of the water, the autopsy report stated, was five feet seven inches tall and weighed 144 pounds. Paisley's Merchant Marine file had him at five feet eleven inches and a robust 170. Was the corpse floating in the water John Paisley?

Why? The Coast Guard reported that on the night of Paisley's disappearance "there was an unusual amount of communication traffic from the Soviet summer residence on the eastern shore of Chesapeake Bay." What was going on? With whom were the radio operators in the waterfront Russian compound at Pioneer Point communicating? Was there an emergency? An intelligence operation in play?

Why? On that same busy night, a Polish merchant ship, the *Franciszek Zubrzycki*, had sailed in the moonlit darkness through the bay on its way to Baltimore Harbor. The next day it headed across the Atlantic to Rotterdam. Had it played a covert role in a spy mission? Had it picked up someone as it made its way through the Chesapeake? Was a secret traveler aboard when it headed across the Atlantic?

Pete thought hard about all these questions. He thought of all the equivocating theories he could present, all the caveats he could raise, all the qualms and qualifiers that any glib counterintelligence officer could reel off. And yet—he knew. He had no doubt.

He's alive: John Paisley had not died that night on the Chesapeake Bay.

He's alive: The gnarled and mangled corpse that was pulled out of the water was not John Paisley. It was not his remains that had been cremated.

He's alive: The walls were closing in, and so Paisley had sent off a prearranged distress signal: Rescue me!

He's alive: A KGB exfiltration squad had run the operation. A bullet shot into someone's head. A corpse wrapped in diving belts. A nameless victim hurled into the sea. A staged suicide.

He's alive: A ladder thrown over the side of a Polish merchant vessel as the ship's engines slow in the midst of its late-night passage across the Chesapeake Bay. From the deck of a small craft floating by the hull, someone grabs the ladder and scurries up under the cover of darkness. He's met on the deck and quickly escorted to a cabin below. He will not emerge from his hideout until the boat docks in Rotterdam.

He's alive: Rather than deal with the public consequences of decades of deception, the embarrassed agency gamely plays along. Paisley was no one of consequence, the press is informed. No documents went missing, no secrets were stolen, no agents have been blown. There are no fingerprint records to identify the corpse. And anyway, it's too late; the body has been cremated.

He's alive: And somewhere in Russia, John Paisley, Moscow Center's long-running mole, was enjoying a hero's welcome.

Or had Pete gotten it all wrong? Had he clutched at clues that led to wild theories? Had he ferreted out sources who were unreliable, who had put a conspiratorial spin on what was ultimately just a sad yet all too human history? Had he, looking to justify a lifetime of his own professional choices, cooked up a case against a poor soul, transforming him into a wily traitor?

After all, he reminded himself reproachfully, there had been a Senate Intelligence Committee inquiry into the circumstances surrounding Paisley's death. Following a two-year investigation the committee had in 1980 decided that they had uncovered "no information which would detract from [Paisley's] record of outstanding performance in faithful service to his country." They had "found no information to support the allegation that Mr. Paisley's death was connected in some way to foreign intelligence or counterintelligence matters."

So was that really that? Was this, then, the final say? Pete, however, couldn't help but notice that this was a carefully worded conclusion; "found no information to support" was a very lawyerly and restrained form of certitude. And it was further tempered by the committee's decision to classify the full report. The details of its investigation into the crucial question of whether Paisley was murdered, committed suicide, or was still alive would, Pete despaired, remain secret, locked up forever in government vaults.

And even its public findings were provocatively undermined by an unlikely yet presumably knowledgeable source—Michael Epstein, the Intelligence Committee counsel who had directed the investigation. "Chances are we will never understand the outcome of the case. It is a mystery. We never really had the resources to . . . investigate it," he said with discernible regret.

The counsel's exasperation was also echoed in the *Washington Post*'s reporting on the secret Senate report. "To many persons familiar with the case," the paper complained, the committee's conclusion was "merely a chapter in a real-life mystery that may never be solved to everyone's satisfaction." And as Pete read on, he found another voice admitting to perplexity. In that same *Washington Post* account, no less an insider authority than the official CIA spokesman had this to say about Paisley's death: "Nobody knows how it happened. We don't know how he died."

But, Pete had to concede, not knowing was, of course, not the same as guilt. Nor was a top secret classification evidence that the agency was deliberately holding back a disgraceful or embarrassing truth. It was the business of intelligent agencies to keep their workings secret. That sort of protectiveness was in their nature. This reticence did not prove anything. The reality was complex, and arguably wrongheaded, but not necessarily sinister.

Still, Pete wasn't the only one whose mind kept suspiciously wandering down these dark alleys. Twelve long years after the *Brillig* had been found floating on the bay, Senator Jesse Helms, a hard-charging conservative senator from North Carolina who served on the powerful Foreign Relations Committee, was posing official questions to his fellow senators about Paisley's grim death. In 1989, he wanted to know whether "John Paisley had been a long-term Soviet mole at the CIA." He wanted the committee to examine "evidence that he might have been recruited." He demanded "the reasons why the U.S. government had waited an inordinate length of time to suspect Mr. Paisley's probable dealings with Soviet bloc intelligence sources." He wanted the committee to request an investigation that would once and for all resolve "whether the CIA had been penetrated by Soviet bloc intelligence sources."

He was stymied. A majority of his fellow committee members voted him down. And the Paisley case, its intertwined secrets and its resonating mysteries, would as a result remain unresolved, troubling theories floating away on an ominous sea of suspicions.

For Pete, it had been a long journey. The relentless sleuth, he had traveled far. From a cozy safe house in Geneva to the venerated plains of Waterloo and on to a defector's dinner party that gave birth to a family of spies. He had set out to avenge agents like Popov, Penkovsky, Trigon, and Shadrin who had paid for other men's sins with their lives. He had fought hard to burst the barriers that still protected Nosenko's secrets. In his old age, a retired spy, he had charged forward to protect the agency. Even when the institution to which he'd devoted his life did not want his help. Even when it lashed out at him, as well as at those with whom he'd been proud to serve. He had refused to retreat despite all the obstacles, all the humiliation hurled his way. And he had done it alone: his duty.

On the other hand, he readily admitted, "I would never get all the answers." He had no doubt that Paisley was the mole he'd been pursuing. But that knowledge was not, he also realized, sufficient. For now all he could do was share his theory with friends, fellow veterans of the Cold War spy wars. Angleton, for one, listened and volunteered that he had no doubts that Paisley was exactly who Pete thought he was. Yet that, too, was only a small comfort.

"I have succeeded in digging out at least the broad outlines of the buried truth," Pete would say in an attempt to find a measure of solace in his tempered success. But he was enough of a realist—and what veteran counterintelligence officer wasn't in his weathered soul?—to grasp that the irrefutable proof he wanted would be forever hidden away in the enemy's vaults along with all its other untold tales. The hunter would need to be able to look deep into the prey's carefully concealed soul before such knowledge could be obtained. But the covert world doesn't work that way. Adversaries hold on to their secrets with all their might. And that would never change.

Part V

"The Other Side of the Moon"

1990–2014

Chapter 32

FIRST THE WALL CAME TUMBLING down. Barbed wire, cinder blocks—all turned to rubble. Then it was as if the world itself was in upheaval. The new push-and-pull economics of *perestroika*. The new accommodating politics of *glasnost*. The Soviet Union Pete had plotted against, the murderous, tribal enemy he had fought tooth and nail, had felt the quivering intimations of its own weary mortality. And by 1990, as the new decade dawned, the Cold War had given way to a Cold Peace.

Pete was astonished. The unraveling of the Soviet Union, so sudden, so unanticipated, took his wandering thoughts back a lifetime earlier. In his schoolboy science class an inviolable truth had been drummed into him. "We will never see the other side of the moon," the high school astronomy text had authoritatively declared. Yet that pronouncement had been discarded to the junk heap of obsolescent science; Pete had grown old enough to see the dark side of the moon prodded, photographed, and mapped.

And this reminiscence got him thinking. At first his musings were vague and mawkish, a senior citizen waxing sentimental about living long enough to see everything. But Pete's thoughts soon hardened. He recalled that after the end of World War II former enemies had sat down together; professional soldiers, warriors bonded by the fraternity of shared experiences, had met to trade combat stories. And he wondered: "If the Cold War was really ending, might KGB veterans loosen up the same way?"

"Their side of old events," he wanted to believe, "could break out some of the buried truth."

His excitement rose at the prospect, but then his confidence abruptly waned and he found himself hesitating. Could he trust his own impulses? Was he just an old man—decades had passed since he'd been in the employ of the CIA—refusing to go gently into the good night? He'd earned his retirement. Why shouldn't he simply sit back in his golden years with his books and his grandchildren and rejoice that his side had won the Cold War? Besides, he'd been down this long and winding road before. He had hacked his way through a jungle of betrayals in a search for the mole, and yet, if he were being brutally objective, all he had to show for his years of exhaustive pursuit was a theory. Walk away, he told himself. Your time has come and gone. The past is past, and one gray-haired retired spy poking into long-forgotten mysteries will not change anything.

But he couldn't quit. The old questions still nagged. As did his sense of duty.

PETE, THEN, BEGAN TO PLOT his new, opportunistic mission. Flying solo, without either backup or, for that matter, the knowledge of any intelligence service, he'd reach out to his fellow ancient warriors who had spent clandestine lifetimes spying for the enemy. He'd try to convince these gray-faced former Soviet Bloc intelligence and counterintelligence officers that the time had come to talk. To sit down, loosen their professional inhibition, and with the liberating spirit of the new openness trade war stories.

But all the time Pete would be the purposeful fieldman. The convivial conversations, the companionable picking through the bones of long-buried operational corpses, would be cover for a more sophisticated mission. Pete wanted the information that would help him at last fully understand what he had previously missed. He needed the firm knowledge that would empower him to once and for all shove away all the lies and myths that the agency, weary and complacent, had decided to accept as truths. He wanted to shake the deep-rooted tree of conventional intelligence wisdom until the mole fell from its branches.

Yet for Pete it was not only about the past. He'd be sounding an alarm. A historical pattern of treachery would be exposed, and with the recognition of new impending danger, a lackadaisical spy service would be mobilized. The warning would be explicit: There's every reason to believe the pattern of treason was still active. New moles had almost certainly burrowed in.

There was also something else goading him on. Although it was not in his nature to articulate this element, Pete was driven by a deeply personal motive. The tirades of abuse that for years had been directed at him, the wounding taunts of "genuine paranoia," of "unsupported suspicions," of "insidious conclusions," all the—he'd nearly bellow—"demonstrable untruths" that had been piled on would crumble under the weight of the irrefutable evidence he'd assemble. His tarnished reputation would be polished, and the now reflexively dismissed legacy of his generation—the accomplishments of patriots like Angleton, Rocca, and Deriabin—would be restored.

And so he went to work. It was laborious, an introduction made here, a letter sent there. Year after year throughout the freewheeling openness of the 1990s until the new century when the Russian Bear started howling with its old menacing growl, Pete was busy making once unimaginable connections. He talked and drank (and drank) with nearly twenty Soviet Bloc intelligence veterans. Under the operational pretexts of conferences or research projects or a documentary in the making, he hurried off to meet with his former adversaries as they passed through Brussels, or Paris, or Berlin. And he took trips deep into what once had been forbidden territory, huddling with old foes in the cooling breezes of the Black Sea resort of Sochi or strolling convivially through Red Square as if unaware of the menacing proximity of the Lubyanka and the sad ghosts haunting its torture chambers.

Not that it was always easy; some wounds never heal. There was the senior KGB general who, teeth bared, admonished, "Remember, we are still working against you." Just as another Moscow Center veteran fixed Pete with an unforgiving stare when he offered an icy reminder: "The KGB is not dead."

But by and large they were surprisingly openhearted, the discussions not sinuous but frank. Veterans who had known the same late-night terrors, who had listened for the heavy tread of footsteps on the stairs or a vengeful pounding on the door, were a community; they had shared common hazards, they had lived lives sharpened by the constancy of similar fears. And so "they seemed pleased and intrigued," Pete marveled, "to talk." Away from the eyes and ears of Moscow Center, they responded spontaneously and with gratifying detail to Pete's insider questions. But they also understood, Pete felt, that he wouldn't misuse their trust. It was never articulated, but a line had been firmly drawn: "I would not ask them to betray their undiscovered spies in the West."

THE DAYS OF PEEKING THROUGH keyholes were over. Instead, the doors were as good as flung wide open. Pete was ecstatic. There he was celebrating in the modest Moscow apartment the state had bestowed on a two-star KGB general for his years of loyal service, a long night fueled by bottle after bottle of vodka, and when the new dawn broke through the window, they were still talking away. Or there was Pete making every effort to hold back his bemused smile as his proud KGB host marched him through the gilded showcase that was the luxurious bathroom complete with golden toilet that had belonged to Victor Abakumov, the infamous Smersh killer and cheerleader for the Great Purge. And there was Pete, his eyes nearly welling with victorious tears, as he stared at the now humbled statute of Feliks Dzerzhinsky, the founding father of the KGB, toppled from his lofty pedestal in front of the Lubyanka. The other side of the moon, indeed.

It was in this deliberate way, gregariously plundering the deep institutional memories of his ex-foes, that Pete learned about the existence of a special army of spies inside Moscow Center. Its operations had been kept secret from the rest of the service. Its ambitious chief was General Oleg Gribanov, the head of the Second Chief Directorate, the KGB division responsible for counterintelligence and internal security within Russia. He had long been a tenacious spy catcher; in the West he'd been known with frosty respect as "the Soviet J. Edgar Hoover." The new covert unit, however, would focus on "operational deception." It was named the Fourteenth Department.

Chapter 33

THE INN WAS IN THE pretty garden village of Prenden, a lush sweep of green lawns and towering copses of trees in the heart of the Wandlitz District that had once been the summer playground of East Germany's Communist Party elites. But by 1994 the country had already been united for four reparatory years, and on a bright spring day like this, Pete believed, it might have been quite easy to forget about its fractious history if it weren't for the cavernous underground complex buried down the road. "Honecker's Bunker," the locals called it in homage to the Communist Party leader who had ordered the construction. Eighty-four thousand tons of reinforced concrete had been poured in preparation for the inevitable day when nuclear missiles would darken the skies and East Germany's National Defense Council, for the greater good of the socialist revolution, of course, would hurry inside to take cover.

This morning, as was his practice, Pete had risen early; in the quiet, he'd try to organize his thoughts for the conference the TV crew would begin filming later in the day. He'd need coffee first, however, and he navigated his way down the coil of twisting wooden stairs to the breakfast room.

Even at this hour the room was bathed in sunlight, a glimmering sweep of starched white tablecloths and heavy silverware. Pete had expected to be alone, but at a small corner table sat another guest. He had an owl's round face and a thick, balding head topped by a halo of wispy gray hair; with his black-framed glasses he might have been an accountant.

Pete recognized him at once. He had never seen him in the flesh, but his photograph had been prominently catalogued in the book of enemy mug shots that Pete had kept stored in his Langley office safe.

Pete did not hesitate. "May I join you?" he asked.

"Please do," said the diner, his English seemingly fluent, as if second nature. He gestured toward the vacant seat to his right.

Introductions were made, but they were all theater. Lieutenant General Sergey Kondrashev of the KGB and Pete Bagley, the former head of the CIA's Soviet Counterintelligence Division, had been tracking each other's footsteps for the past thirty years.

The orange juice was freshly squeezed, the coffee strong, and the conversation flowed. As spies often do, they reminisced about how they'd gotten into the Great Game. Pete offered a portrait of his family's naval history, certain he was sharing nothing that the Russian didn't already know. His weak eyes had forced him to look for another line of work, and, lo and behold, he'd wound up in the CIA, he said good-humoredly. The Russian made a show of being amused, too.

When it was Kondrashev's turn, he casually confided that he'd started out after the war in the internal-counterintelligence directorate. For many years, his primary target had been the American embassy in Moscow.

That got Pete's interest; and like an unconsummated romance that had remained tantalizing despite all the passing years, an old case suddenly came alive. Its central figure had been a snarling KGB hood who'd been working against the American embassy in Moscow during Kondrashev's tenure. The agent had made a sudden trip to Washington, and for decades Pete had wondered why. But should he ask now? A single rash question could demolish Pete's long-term mission in an instant. A disconcerting vision of an enraged Kondrashev jumping to his feet and stomping off, never to speak to him again, flitted through his head. Tread gently, Pete cautioned himself.

But the moment was too promising to waste. With a blitheness that was all disguise, smiling ferociously for additional cover, Pete, without either preamble or explanation, boldly announced that there was "something that's been bothering me for a long time." Why, he asked, had an operative whose usual target was the Moscow embassy arrived with a new name, a new cover identity, for a brief visit to Washington? What, Pete inquired pointedly, had your colleague been up to in America? Why had he hustled off from his mainstream duties for a quick excursion to the United States?

Then he waited. As the heavy silence dragged on, Pete grew anxious and chastised himself. His tradecraft had been deplorable; he was indeed getting old. He should have charmed. He should have flattered. He should have built up to the moment. Or maybe he should have kept his suspicions to himself. Now the best he could hope for would be a noncommittal shrug. Or let Kondrashev plead ignorance; at least that'd save face for both of us. It would allow the relationship to continue, and perhaps one day down the road even blossom.

When the Russian finally responded, his directness mirrored Pete's. "Oh, that was to meet an important agent," he said flatly. Then he paused, perhaps debating whether to go on. "One who was never uncovered," he added at last.

Nothing in Pete's deportment suggested that a long-gestating suspicion had been confirmed. And he knew better than to prod further; he would not jeopardize the future of this extraordinary relationship by trying to get Kondrashev to reveal the identity of the still-unknown spy the operative had contacted. "Yes, I've long thought so," Pete replied. And with that, he let the matter drop.

But by the parsimonious standards of the secret world where

secrets are capital never to be spent, both men recognized that it had been an exceptional exchange. And a groundbreaking one: a harbinger of what was to come.

Over the next thirteen years, until Kondrashev's fatal heart attack in 2007, the two old professionals, in time friends, kept in close touch. And in the course of their productive collaboration buried secrets were exhumed, and old suspicions were laid to rest. With Kondrashev pointing the way, and with Pete collecting new leads and confirming evidence from a growing network of former Soviet intelligence officers, he found the answers to the questions that had been haunting him for years.

Who told him what? How did a specific scrap of information fit into the puzzle he was assembling? Who were his sub-sources? This was intelligence that Pete, with a professional's concern for his Joes, refused to reveal. And with the passing years, he recognized the ominous portents building in Putin's new Russia. "Secrets," he saw with distress, "are more severely defined and more jealously guarded than during that short period I enjoyed after the Soviet collapse." He feared that if he identified the members in his group, there would be retribution; they'd be stripped of honors and pensions, and even after their deaths their families would continue to pay the heavy price levied by a vindictive state.

Pete decided it'd be prudent to use Kondrashev, who had been boldly open about his relationship with the former CIA officer to the new breed of spymasters (foreign operations now run by the SVR; internal security handled by the FSB), as the spokesperson for the group of sub-sources. He'd be the public stand-in for the disparate voices of about twenty knowledgeable former KGB intelligence bigwigs who over several years had helped shape Pete's understanding.

Therefore, when Pete sat down to write the book that would finally reveal the story of the clandestine Fourteenth Department and how this discovery had put him on the path to solving the mysteries that had been tormenting him for decades, it was a tale told through Kondrashev's solitary voice. And this was largely accurate; his words had been the loudest, the most forthcoming. But a chorus of professionals had chimed in, too, and their earnest contributions had also helped to bring the hidden history to operational life.

LIKE PETE, KONDRASHEV HAD A historian's fascination for antecedents, and so rather than plunging right into the workings of the Fourteenth Department he started with Operation Boomerang.

In November 1958, he began—and as in all things, Kondrashev was punctilious with chronology—GRU Lieutenant Colonel Pyotr Popov had been called back from his post in East Berlin to discuss Moscow Center's concerns over the recent flurry of student protests. But the stated reason had been a ruse. No sooner had he arrived than he was hauled down to the dank cellars of the Lubyanka for a decidedly hostile interrogation.

The singular truth of pressure, Kondrashev went on, is that everyone breaks; it's only a question of how long and arduous the ordeal will be. Popov's confession came in bits and pieces. He acknowledged "errors of judgment" but he struggled to hold on to the large secret he was concealing. Inevitably, though, he revealed everything: For six years he'd been delivering a steady stream of official documents detailing Soviet weapon and nuclear warfare advances to the CIA.

As Kondrashev spoke, Pete's memory filled in other more affecting details. He saw himself in his days as a young fieldman in Vienna. A bone-chilling night, stomping his feet in a futile effort to keep warm, as he waited in a doorway on a dark, narrow street near the center of the city. His service revolver lay heavy in his jacket's outside pocket, within easy reach, and his eyes were peeled for the enemy. Across the way there was Popov, bundled up in a dark blue overcoat, walking briskly down the street to make contact with his handler. Pete remembered his absolute elation, the heady sense of victory, when later that night the package Popov had delivered was reviewed: Field Service Regulations of the Soviet army, the enemy's operational bible the Pentagon had been desperately trying to obtain. For three tense but incredibly productive years, Pete had been part of the four-man team running Popov. And he recalled the devastating sense of loss, the sort of deep grieving and bewilderment one suffers when a friend dies, when he had learned that Popov had been executed.

It was that memory which drew Pete abruptly back into present. He realized what he'd just been told.

November 1958, he interrupted. You said that was when they broke Popov. But he was still working for us for nearly another year. He wasn't executed till, what was it, the end of October 1959?

Kondrashev offered an indulgent smile. Exactly, he agreed. Now if you'll let me continue, I will explain.

After Popov's confession, the Russian resumed, Moscow Center became a very gloomy place. Our agents had been blown. Our operations revealed. We were bleeding. And all the brass could do about it was fire the GRU chief, General Shalin. Which of course was too little, too late.

Enter General Oleg Gribanov. "Gribanov was a ball of fire," Kondrashev told Pete, growing more animated as he moved into the heart of his story. He saw that there was a way out of the wreckage. Since its early days, *nastupatelnost*—aggressive counterintelligence—had been the KGB's "guiding principle." Gribanov realized he could employ the trove of GRU secrets Popov had exposed as a weapon against American intelligence. He called this new mission Operation Boomerang.

To set Boomerang in motion, he created the Fourteenth Department. The operational concept was simple yet inspired: recycle the now worthless information Popov had revealed, throw in a scattering of new secrets that seemed substantial but in reality were just chickenfeed, and then play it back to the enemy. Yet all the time, the Fourteenth Department would be leading the CIA down blind alleys. And, more significantly, the manufactured intelligence would hide the scent of Moscow Center's big secret—how Popov had been discovered.

Pete listened with attention. His gaze remained as direct and focused as before. And nothing in his demeanor gave a sign that Kondrashev had alluded to precisely what he'd long suspected: the existence of a mole inside the CIA.

It took all Pete's discipline, he'd later confide, not to interrupt. His entire being was imploring him to pose a single burning question. Give me a name! he wanted to shout. But would that question ruin everything? Would it bring the entire conversation to a screeching halt? Would Kondrashev, a man who'd devoted his life to serving his country, come out and docilely reveal his service's greatest Cold War secret?

Patience, Pete decided. He would play this game long. He judged it would be wiser to let the Russian take the lead, to let the conversation unfold according to Kondrashev's orderly plan. But when the moment was right, when there was nothing left to lose, he would ask his question. Hell, he'd scream it at the top of his lungs. He would get the name of the traitor. He just needed to wait.

In the trade as in life, the professionals say, timing is everything.

AND IF KONDRASHEV HAD ANY idea of the rebellious thoughts Pete was reining in, he gave no indication. Instead, his narrative veered leisurely backward, returning to focus on a discrepancy Pete had previously spotted.

You were correct, he said matter-of-factly. Popov was arrested in 1959. But I was also right: he had indeed confessed a year earlier. Popov, you see, was the Fourteenth Department's first case. Gribanov had doubled him back against you. He ran Popov for a year. The documents Popov delivered from his new post in Moscow were counterfeit. Worthless inventions written by a mischievous Fourteenth Department team. They'd been crafted to send the CIA running in circles. At the same time, Operation Boomerang was also a learning experience for Gribanov and his secret army. An opportunity for them to perfect their skills.

"And don't believe for a minute," the Russian suddenly went on, according to the account Pete would write of the conversation, "that old story that we detected Penkovsky by surveillance."

Pete knew he was referring to GRU Colonel Oleg Penkovsky. The history books lauded him as the greatest spy of the Cold War. The double agent whose daring had "saved the world" during the 1962 Cuban Missile Crisis. And it was a case that had also fueled Pete's suspicions.

The truth was, Kondrashev continued, we were on to him for a while. The Fourteenth Department even took the risk of letting him go abroad. Why? Even now, he said emphatically, that cannot be disclosed.

Yet Pete suspected that he knew. Moscow Center had delayed Popov's arrest to disguise the fact that he'd been betrayed. The twelve months had given them time to lay credible groundwork for a cover story. They concocted a boastful yarn about Popov's being blown because of the dogged skills of the KGB watchers. And the CIA fell for it. And since it'd worked once, the Russians figured they'd play the same trick again. And it succeeded again. The KGB dusted off the same lie to explain how they tumbled onto Penkovsky.

But both of the double agents had been revealed by a mole.

Kondrashev, Pete felt with mounting excitement, was taking him closer and closer to where he wanted to go. He could feel the heat of the big secret that lay buried beneath the years of deception. But for now all he could do was wait.

It was evening, the end of a long day, and Pete and Kondrashev were walking side by side around a large pond. In the fading light, croaking bullfrogs were making a surprisingly ferocious racket. Both men were bone tired, and it was a while before either spoke.

Earlier they had stopped for a drink, and with mutual pride they'd shared photos of their grandchildren. Pete imagined the conversation would continue in that benign direction, into family matters.

Kondrashev returned instead to the Fourteenth Department.

Now that Gribanov had succeeded in sending disinformation into the West, Kondrashev said, picking up the story as if there had never been a pause, he initiated the next stage of his operation. He gave the enemy a spy.

By "give," Pete understood, the Russian meant "dispatched." The Fourteenth Department had sent to a double agent to the West.

"I was the only person in the whole First Chief Directorate who even knew of this operation's existence."

"How was that managed?" Pete asked. He had a professional's curiosity in how the opposition went about its work.

A Fourteenth Department officer, Kondrashev explained, would regularly show up at his office with a stiff leather portfolio. Inside would be a progress report on the operation. The courier hovered as Kondrashev read; Gribanov had given his man strict orders not to let the report out of his sight. When Kondrashev finished, he'd sign a form acknowledging that he'd been made privy to the case, and then he'd hand the report and the form to the courier. Both documents would go back into the leather portfolio, and then the Fourteenth Department officer would leave. No one else in the First Chief Directorate ever saw the reports. Gribanov, Kondrashev volunteered, was fanatical about keeping secrets.

"Why you?" Pete asked.

With a small burst of vanity, Kondrashev curtly reminded Pete that he'd been the deputy chief of Foreign Intelligence; he had to be apprised of every overseas disinformation operation. And not even Gribanov was prepared to lock horns with the generals who had made that iron rule.

"It was Polyakov," the Russian said, suddenly revealing the name of the first agent the Fourteenth Department had dispatched.

"What?" Pete exclaimed.

But he wasn't confused. He knew all about Dimitry Polyakov, a lieutenant colonel in the GRU. He had delivered secrets to the FBI when he'd been based in New York, and then after he'd moved on to Rangoon, Moscow, and later for two separate tours in New Delhi, he'd been run by the CIA. When he was caught by the KGB and executed, the agency went out of their way to praise him publicly as their "greatest source of the Cold War."

And for forty years Pete had been convinced that Polyakov had been a KGB plant.

Ridiculous, the agency had shot back. Stop seeing plots everywhere, the fifth floor had sternly lectured. No intelligence service would hand the opposition a spy from within their own ranks. That kind of risk was unthinkable. And Pete knew the institutional mindset had never changed. The operating rule today was that if a walk-in was a Soviet intelligence officer, then put out the welcome mat. He had to be a genuine defector.

Except Kondrashev was now confirming that the Fourteenth Department had been set up to do precisely that sort of trickery: to dispatch staff officers as doubles.

But Pete wanted more. Taking refuge in a perplexed tone, he played devil's advocate; at the same shrewd time he hoped the challenge would push Kondrashev into a further revelation.

"But they executed Polyakov!" he said. "Why would the KGB execute a man who they themselves had sent out to commit treason?"

"Because they found out he was giving you more than he was supposed to."

"Found out? How?" Pete bluntly demanded, and no sooner had he spoken than he worried he'd been reckless.

Kondrashev paused. He seemed very much aware that the conversation had moved into forbidden territory. He considered his words carefully, and when he spoke his voice was firm. "They found out through a source inside American intelligence," he said.

Pete was rocked. He tried, however, to remain impassive. He

was determined not to betray his exhilaration. But he didn't have the actor's skill. Kondrashev seemed to sense his new alertness. And he could feel the Russian's growing discomfort. At that combustible juncture, all Pete's instincts told him that Kondrashev had gone as far as he would for now. It was even possible that he'd revealed more than he'd planned. Pete backed off.

Locked in a companionable silence, they headed off for a nightcap. But all the time Pete's mind was shouting the cathartic words he had just heard: *They found out through a source inside American intelligence.* From a mole.

AND IN TIME, NURTURED BY patience and coaxed on by their growing friendship, the trail led, as Pete had no doubt it inevitably would, to Nosenko.

Yet there was no thunderous drum roll. No trumpet sounded to signify that a great revelation was at hand. Kondrashev had launched into what initially struck a disinterested Pete as simply a discourse about the treacherous intramural sport of office politics. Anticipating the punch line, Pete expected to hear that in Moscow Center, as in Langley, a careful spy had better watch his back.

In early February 1962, Kondrashev had begun, he'd been ordered to report to the office of General Gribanov. He was surprised and, to be honest, a bit wary. The way things worked in the KGB, you never knew if you were going to get a medal or be dragged off to the basement in chains.

When he arrived, Gribanov got right to the point. The Fourteenth Department, he said, was preparing to launch a complex operation against the CIA. He wanted Kondrashev to help him run it. Leave the First Chief Directorate, he offered, and come work for me. He'd make him his deputy. And, a further enticement, with the new job would come a new rank. Kondrashev would be promoted to a one-star general.

A lot of things went through Kondrashev's mind at that moment. The Fourteenth Department was rumored to be involved in all sorts of daring and complex missions. And the general's rank would be a boyhood ambition finally realized. As well as a raise in pay. Possibly even a bigger apartment. But he told Gribanov that he needed a day to think. It would be disloyal to walk off from his present job without first having a discussion with General Ivan Agayants, his boss.

What's there to discuss? Gribanov challenged; brusqueness was his style. And he wasn't concerned about offending Agayants by poaching his deputy. Kondrashev, however, refused to be bullied. He repeated that he needed to talk to Agayants before making a decision. He owed him that courtesy.

"I understand that you're tempted," Agayants said when they met. "But relax. You'll get general's rank soon enough in the First Chief Directorate. And in this particular operation, you wouldn't be doing yourself any good in the long run. Gribanov is going to screw it up. He's rash, doesn't have time for the detailed preparations that these things need."

Kondrashev decided to listen to his boss. He turned down Gribanov's offer.

Pete assumed this was the end of the story. But he was mistaken.

Kondrashev revealed the operation the Fourteenth Department had wanted him to run. It was to help launch a Second Chief Directorate officer who would volunteer to spy for the CIA. He would approach the Americans in Geneva, where he'd be working ostensibly as a watchdog for the Soviet delegation at an arms conference. His name was Yuri Nosenko. His mission was to protect the source we had in America.

At that gratifying moment, Pete remained quiet, even immobile. He had the dizzying feeling that his life had come full circle: from that safe house in Geneva to the years of relentless pursuit without the achievement of conclusive results to his present understanding. The Fourteenth Department operations had been launched with a single horrifying objective: to protect the mole who was hidden at the end of the long dark tunnel. All the cruel deaths, the slaughtered agents from Popov to Penkovsky to Trigon and to Shadrin, had been victims, in part or wholly, of this traitor embedded in the CIA. The double agent Nosenko had been dispatched to protect.

The agency had refused to listen to Pete's warning. Instead, they'd hampered his hunt, mocked his suspicions. Sneered that he was paranoid, obsessed. But for the first time Pete had his vindication. For the first time he had the confirming proof he'd always needed. The evidence he'd felt honor bound to find.

And as if reading Pete's mind, Kondrashev spoke up. "How could your service ever have believed in that man? How could they have accepted Nosenko?"

ALL THAT WAS LEFT, THEN, was one final question. One question whose answer would lead his long hunt to a satisfying conclusion. One question whose answer would clarify the only mystery that still remained.

Pete turned to Kondrashev and asked, Who was John Paisley?

Chapter 34

AS THEY APPROACHED, THE SNOW had begun falling heavily from the leaden Moscow sky. It swirled about, and it was hard to see very far into the distance. Pete thought the red granite outbuilding might be the gatehouse of a church, or perhaps it was a companion to the cluster of baroque buildings and domed towers that made up the sixteenth-century cloister they'd just toured. He turned to Kondrashev, but his look was ignored; his friend had been acting in that disarming way all morning, at times terse, at others uncommunicative to the point of rudeness. But Pete was in no position to complain. After the way he had behaved last night, Pete feared he'd strained their friendship to the breaking point. Kondrashev, Pete could not forget, had nearly bawled him out. And he'd deserved it. Still cringing, Pete conceded it'd been all his fault.

It had started over dinner in Kondrashev's small apartment just across the river from the Kremlin. Pete had arrived with a bottle of wine in his hand, and an agenda locked firmly in his mind.

What next transpired was nothing less than an interrogation, and it commenced before the first glass of wine had been poured and continued relentlessly through dinner. Kondrashev had tried to deflect, to twist the conversation into other more genial realms, but Pete would not be deterred. Old spies understand their missions, like their days, are numbered.

Tell me about Paisley, Pete attacked, according to the account he later shared with a friend.

Tell me, was he the mole? Were all the Fourteenth Department plots launched to protect him?

I only know what I read, Kondrashev parried without conviction. The news reports said he'd committed suicide.

It wasn't suicide, Pete shot back. It *certainly* wasn't suicide. It's not even a good cover story. It's simply a lie. Tell me what you know, he repeated.

I don't like these questions, Kondrashev said testily. Badgering me in my own house. From a friend. It's not proper.

So what's the answer? Pete continued, ignoring the rebuke.

What answer? Kondrashev shrugged.

What do you know about Paisley? You saw the Fourteenth Department operational reports. You told me that. You said it was your job to read every single one of them. So you know. You know who was Nosenko's control. Who he'd been instructed to run to in case of emergency. Who he'd turn to to put the fires out.

You are asking me to betray my country. To divulge state secrets, Kondrashev said, his voice at last rising in anger.

Maybe I am, Pete agreed. But he was not backing down.

This is an old ghost you're chasing, Kondrashev implored. Let him die in peace. Don't you understand? You're not responsible anymore. Neither of us are. Our time has come and gone. Let's open another bottle. Forget about Paisley.

I can't, Pete said. He spoke with the perfect calm of a man stating an irreducible truth.

The conviction in Pete's voice appeared to leave Kondrashev flustered. He needed, it seemed, to gather his thoughts, and when he finally spoke his tone was brisk and resolute. I suddenly find I'm not feeling very well, he said. I need to get some rest. And you need to leave.

Pete had been surprised when Kondrashev called the next morning and suggested they meet. The Russian said he'd be in the hotel lobby in twenty minutes. The invitation had been clipped, even frosty, but Pete anticipated that over breakfast Kondrashev would warm. And Pete suspected that he'd need to offer some sort of apology, or at least try to justify his belligerence.

But there'd been no breakfast. Instead Kondrashev, without much of an explanation, led Pete to the metro. At the Luzhniki Stadium stop, the Russian exited and Pete followed in mute obedience. It was a short walk in the cold, and then, with Kondrashev still leading the way, they passed through the high masonry walls of the Novodevichy Convent.

It was a fantasyland. Scattered about an expansive park were a dozen or so baroque buildings and crenellated towers, a gilded imperial refuge where the Russian royal families had exiled their maiden daughters who had been forced to take the veil. Pete wandered about fascinated; the experience of making new discoveries, of learning something he hadn't previously known, had always been one of his great pleasures. And no less gratifying, he suspected this was a carefully considered attempt to mend their broken friendship. Kondrashev was giving him a gift.

But all the while, Pete later decided, Kondrashev had been finding the strength to firm up his decision. It had been a tactical delay. He'd been trying to muster the will to open the last important lock.

They'd been looking up at the five golden domes of Smolensky Cathedral, the church spires rising up into the darkening winter sky like the turrets of a gloomy castle, when Kondrashev ordered, Come, let's go.

They'd exited through an opening in the convent's southern wall, and a minute or two later they were approaching the red granite gatehouse.

As Pete now got closer, the building, despite its fresh dusting of snow, remained dour, resolutely grim. Adjacent to it, a curving brick wall, utilitarian rather that decorative, encircled what must have been acres and acres of flat terrain. And he saw the orderly rows of gray gravestones.

The Novodevichy Cemetery, Kondrashev explained.

And the last stop in Pete's long journey.

It was seven years later, and Pete was dying. He'd been battling cancer for four years, ever since it had been diagnosed by an apologetic doctor in Belgium back in 2010, but now the fight was coming to an exhausted end. He was in the home he loved, his wife and family—a daughter and son-in-law who lived in the covert world, another son and daughter who had chosen the overt world—at his bedside, and to all he seemed neither fearful nor bitter. "At peace," was how one friend put it.

In the weeks before his death, he'd been asked whether there was anything he needed to do, any final words he wanted to share. Pete thought about that, and despite tormenting spasms of pain, he found the discipline to tie up a few loose ends.

He wrote about what he had uncovered as he fought his way through the legacy of betrayals that had undermined the agency. He wanted his words to be an indictment of institutional negligence, of professional doctrines that had failed. And he wanted his words to be a reminder that traitors continue to threaten America's intelligence services.

Yet as much as he might've wanted to, he could not, he explained to a friend, bring himself to write about what he'd learned that day at the Novodevichy Cemetery. During his career, he'd felt the sting of Langley's vipers. And they had shown they were unforgiving, still ready to strike. The agency's long knives, he had remarked to several acquaintances, remained steely sharp, poised to slice vindictively away at his reputation. He did not, as these friends remembered the conversations, share this observation with bitterness. Rather, it was with a tone of resignation, the way one learns to accept an incurable illness.

Nevertheless, Pete was well aware of the vigorous case that was being made against him as he had continued to speak his mind. An essay published in the official agency journal offered a stern verdict: "Bagley identified a plausible motive for Soviet deception and supported it with voluminous circumstantial evidence, yet Nosenko was not under Soviet control." And a subsequent article in the same authorized journal went even further, its taunt implying that he was simply making things up as he went along. The writer snidely noted "Bagley's reliance on unnamed former KGB officers as sources for essential (some would say convenient) information."

Things, in fact, had gotten very personal. There had been, for one example, the time he, a retired officer, had been set to give a presentation at the CIA auditorium, and at the last minute the powers made sure it was canceled. And for still another petty example, they'd also intrigued to cancel his talk at a downtown Washington museum.

Only it wasn't just him, Pete feared, they'd come gunning for if he disclosed all he'd come to believe. The alleged sins of the father, he'd recently discovered to his dismay, were passed on in their merciless judgment to the children.

In March 2012, the Wilson Center at Georgetown University had put together a symposium on "Moles, Defectors, and Deceptions." The conference had the agency's tacit blessing; several of the participants were either past or present CIA employees and the agency's press office eagerly distributed transcripts to reporters. Pete had been asked to make a presentation over the phone from Brussels. He spoke candidly and his views were pretty much dismissed ("I would hope that you all have a healthy skepticism for former KGB officers telling the truth to Tennent Bagley," urged one longtime CIA officer), but Pete was used to this. Besides, he equably acknowledged, differences of opinion were part of the process that wove together the fabric of counterintelligence. But what had blindsided him, left him seething in fact, were the caustic, disquieting barbs hurled by one of the participants.

Carl Colby was the documentary filmmaker son of CIA director William Colby, the man who had forced out Angleton and Rocca. The director who had reinstated Nosenko on the agency payroll and had dismissed the hunt for a mole as "counter-productive." And now his son, although not an agency employee, was back in the trenches fighting an old war at the conference against a new enemy.

"As a final little note," the CIA director's son asked as he concluded his presentation, "who are my father's descendants, his progeny? Who are we? Who are the Colbys? Well, there's an international banker. There's an attorney at the Department of Justice. . . . There's me. There's my son, a US Marine officer. There's an analyst who works on nuclear disarmament. And there's an educational consultant."

"And Angleton's daughters?" he nearly sneered. "They're both Sikhs living the high life in a temple complex in Santa Fe."

Sikhs living the high life. Pete had no doubts; the opposition was a vengeful lot. He didn't want his final words to ignite a taunting, disparaging battle that would continue after he'd left the field of combat and could no longer defend himself. It would be an argument he could never win. More disconcerting, there was no telling who else would suffer under their attack.

He was satisfied. He *knew,* he confided to two friends. Yet he also understood the indefinite quality of the proof he'd found at the Novodevichy Cemetery would never sway unreasonable, predetermined minds. He could almost hear the contemptuous dismissals of his "unnamed sources" and "convenient information." After a lifetime of secrets, he resolved to go to his grave guarding one more.

And so one night Pete, weak and growing closer to his end, was with his family in the comforting cocoon of his wood-paneled study. He had asked for the video of the BBC production of *Henry V* to be played. It was his favorite Shakespeare play, and he knew the words to many of the scenes by heart. His habit had been to speak along with the actors, but tonight he couldn't seem to summon the energy.

Still, when David Gwillim, the actor who so affectingly played the valiant English king, began his rousing St. Crispin's Day speech, Pete tried to mouth the text, too. At first it was a struggle, but he fought on, willing himself to persevere, and his words grew precise and clear, the lines spoken with an ardent conviction:

"Old men forget; yet all shall be forgot.
But he'll remember, with advantages,
What feats he did that day. . . ."

And at that moment, recalling all that lay behind and staring into all that lay ahead, Pete might very well have thought, Yes, that's me.

Pete died the next day.

AND WHEN THEY HAD STOOD outside the entrance of the Novodevichy Cemetery, Pete had turned to Kondrashev and asked, So what are you saying?

"I'm saying that there are some things I can't say. That I can never say." The Russian's tone was neither conceding nor confiding.

And if I looked? Would I find a grave?

First you might need to find a name, Kondrashev suggested. It was the Fourteenth Department's practice to bury its deceased foreign agents using cover identities. That way, he said, secrets remain secrets. Intrigues can continue. He spoke as if he was talking about something abstract, something not at all connected to his reason for bringing Pete to this cemetery.

Pete retreated into a ruminative silence. He considered what he had come to learn today, and what he still didn't know. He weighed the two conflicting enigmas in his mind, and in the end he was left with a powerful sense of satisfaction for what he'd achieved.

"It's good to understand things, even if you can't do anything about them," he said at last, according to the account he'd shared with friends.

"My friend, that's what old age is all about," said the Russian with a small laugh.

Pete smiled, and freed from the intensity of his thoughts, he noticed that the snow had started to fall with a sudden fury. And so the two old spies went off in search of someplace warm, where there'd be a fire blazing, perhaps even a bottle of vodka to share at the end of their journey.

Epilogue: The Weight of Secrets

A SECRET IS A BURDEN. THE weight is crushing. And Maryann Paisley's secret, sinister and leaden, pressed down hard. Unremitting. Unbearable. She needed to confide in someone.

Confession, she told herself, would not be a betrayal. It would be cathartic. But even as she told herself this, it still felt wrong. She could not do that to her husband in spite of all he'd done.

Instead, Maryann devised a new strategy. She'd provoke them. Force them into action. Their profession was secrets. If she attacked, they'd have no choice but to come to her rescue. They'd lift this millstone. They'd help her.

She wrote to the CIA.

"I find writing this letter a difficult but necessary task," she began. "Throughout my twenty years as a CIA wife, I have felt that I could depend upon the Agency to help me anytime that my husband was not available." Only now there had been "a betrayal."

She hurled their lies back at them. Her husband's activities "were certainly not confined to the overt side." The autopsy had been a charade; the height and weight of the body pulled from the bay had no resemblance to John Paisley's. The CIA's and FBI's claims that there were no fingerprint records for her husband were ludicrous. And why, dear God, had there been such a rush to have the body cremated? She hadn't even had a chance to see the corpse.

Her concluding charge: She did not believe it was John Paisley who had been cremated.

It had felt good to say that. Yet she had not told them everything. She had protected her secret. But she had shared enough. Now she waited to see what they would do with it.

The initial response left her enraged. The letter came from CIA director Stansfield Turner. The agency, he wrote, had no investigative powers and "must defer to the Maryland State Police."

And then she grew frightened. Her lawyer discovered a tap on her telephone. He believed there were listening devices planted elsewhere in her home. After consulting with electronic specialists, he charged that a sensor had been installed in the chimney to send a signal whenever someone entered or left the home.

After that, things only got worse. Out of the blue, James Angleton called and invited her to lunch. He said he was retired, but she knew there was more to the story: he'd been forced out of the agency. But she went anyway. She was desperate. The secret was killing her. Perhaps he could take on some of her burden. Perhaps the famous spymaster could ease her pain.

They met at the Army and Navy Club in downtown Washington. They drank martinis. He asked lots of questions about her husband's career, about his overseas travel. And when he was done, he told her that he believed her husband was a traitor. That at some point in his life, John Paisley had decided to work for the Soviet Union. What do you think? he challenged.

Maryann did not know what to think. So she excused herself and left. And she took her secret with her. She would continue to endure its oppressive weight rather than give Angleton, or any of them for that the matter, the prospect of revenge.

And so she did not tell anyone about the postcards that had been received. A different one every month. One was posted from Valparaiso, Chile. "How is everybody? How is the family? Hope to see you?" It was signed "Sandy," and the only Sandy she knew was a friend of John's who had disappeared when his flight to Sri

Lanka had vanished from the radar screens; neither the plane nor any bodies had been found. In another card, "Sandy" quoted the final lines from John Masefield's "Sea-Fever." It had been her husband's favorite poem, and he knew the words by heart:

"And all I ask is a merry yarn from a laughing fellow-rover,
And quiet sleep and a sweet dream when the long trick's over."

It was nearly destroying her, but she decided to keep her secret close to her heart. The long trick was over, and now she would grant him his quiet sleep. She would not tell anyone that she knew John Paisley was still alive.

A Note on Sources

When Pete Bagley, the hero of this story, died, the obituaries made quick work of summing up what had been a varied and complicated public life. "Played a key role in the controversial handling of Soviet defector Yuri Nosenko," was how the *Washington Post*'s lede encapsulated Bagley's activities. The *New York Times*' opening graf stuck to this territory, too, albeit in a more nuanced fashion: "a former CIA officer who helped a mysterious Soviet spy betray his country, then tried for a half century to prove that the defector was actually a Russian double agent."

Yet while the Nosenko affair was indeed central to Bagley's professional life (the "Rosetta stone," he called it), I began my exploration of his remarkable career with the intent of focusing on what had struck me as a more important truth: how the case opened his detective's mind to the belief that the CIA had been penetrated by a mole. And as I set out to tell this story, to recount the perils, pitfalls, and ultimate success of Bagley's long-running mole hunt, I was also prodded by an observation Ed Epstein, a groundbreaking investigator into the secret workings of the US intelligence community, had offered on his friend Bagley's quest. "How he found the answers on his own could provide the plot of a great Hollywood spy movie," Epstein had provocatively written—before his essay, to my frustration, had quickly moved on to other matters.

From the outset, therefore, I was guided and encouraged by two ambitions. I would tell the story of a real-life pursuit of a traitor. And I also was determined to shape this tale as a nonfiction narrative mirroring the actual adventure Bagley had lived.

Yet as I proceeded, I discovered, to my increasing consternation, that I had entered an investigative minefield. At every stage of my inquiries, I encountered a good deal of resistance. There were seemingly knowledgeable individuals in the covert world who for a variety of deeply held reasons—some intent on protecting at all costs the reputations of the institutions they'd served, others bristling with surprisingly durable personal antagonisms—refused to engage with the reality of the events that I'd uncovered and shared with them. Battle lines, apparently, had long ago been drawn and with the passing years had become reified. And in this grudge war (a taste of which I try to give in my tale), the truth—and its crucial implications—became in many quarters an irrelevancy. The official mindset was, in effect, to let sleeping moles lie.

And no less an obstacle to a writer trying to get to the bottom of things with some authority, sources who were (only after considerable prodding in most cases) willing to talk were nevertheless reluctant to allow themselves to be identified. Time after time, they shared critical, previously unreported information, yet they were adamant that their names could not be used. Part of their logic was professional: spies, they felt, should remain in the shadows. Yet another large component, I discovered with dismay, was their fear of reprisals; character assassination, as my account suggests, was an often-deployed weapon in the spy vs. spy wars that to this day rage within our intelligence services. And this apprehension also affected the friends and family of both Pete Bagley and John Paisley. I talked to several of them at length, and still they acceded to these interviews (many spread across multiple days) only if I promised that I would not identify them.

It is an agreement that I am honor bound to keep.

And yet this book suggests some startling new truths.

So how did go I about getting to the bottom of things? How

did I manage to take the reader on a journey that culminates on a snowy afternoon at the entrance to an ancient, venerable cemetery in Moscow? And how can I satisfy the reader that (as an in-house CIA journal sniffed about Bagley) while some of my sources are "conveniently unnamed," this is a true story?

And, no less of a challenge, how did I craft a narrative that tries to have the intrigue of a mystery and the momentum of a thriller—while also being a true story? Specifically, how did I accomplish this without resorting to a sputtering narrative, one that tediously reiterates the sources underlying the highly charged drama shaping each incident I recount? That doesn't trudge on like an academic tome?

Here, then, are the cardinal rules that guided me as I wrote this story: If a statement is in direct quotes, it is information that was conveyed to me in that precise form in an interview, a government document, a published book, or a press report. And if an incident is depicted, its details were shared directly to me by at least two mutually confirming sources, or substantiated in government documents or previously published accounts.

Consider, for example, the sections on Maryann Paisley that bracket the gist of the narrative. She died years before I began my research; I did not interview her. Her thoughts and opinions, however, were conveyed to me by members of her family, documents obtained by the Freedom of Information Act, interviews with her friends, interviews with individuals who had spoken with James Angleton in the months before his death and had knowledge of his luncheon with Mrs. Paisley as well as his long-gestating beliefs about the significance of the Paisley case, lawyers' briefs filed on Mrs. Paisley's behalf against the CIA and Justice Department, the transcripts of the insurance trial after the car crash involving her son and the death of a passenger in the car he'd been driving, and statements previously published in books and newspaper reports.

In the course of my research for the entire book, I conducted eighty-three separate interviews, including several that were quite lengthy. I also relied on many recently declassified government documents, including, for example, the FBI file on Yuri Nosenko (File Number 63-68530) that ran to 718 pages; CIA files on Pyotr Deriabin, especially those newly declassified accounts of his behind-closed-doors testimony to the Warren Commission; and hundreds of pages of Freedom of Information documents on the John Paisley case that had been originally requested by members of the Paisley family and their lawyers.

Also invaluable were the convincing firsthand accounts Pete Bagley had written (*Spy Wars* and *Spymaster*); the many books and articles on the Paisley case (particularly *Widows* by William R. Corson, Susan B. Trento, and Joseph J. Trento, and the investigative reports in the *New York Times* by Tad Szulc and William Safire); and a tall mountain of books on the CIA mole hunt (most helpfully *Wilderness of Mirrors* by David Martin; *Molehunt*, by David Wise; *The Ghost*, by Jefferson Morley; *The Secrets of the FBI*, by Ron Kessler; and *Angleton Was Right*, by Edward J. Epstein).

What follows are the principal sources for each chapter of this book.

Prologue: William R. Corson, Susan B. Trento, and Joseph Trento, *Widows* (New York: Crown Publishers, 1989) [*Widows*]; Interviews with Paisley Family Sources [Paisley]; *Maryann Paisley v. The Travelers Insurance Company* Depositions [Paisley Depositions]; Fairfax County, Virginia, Courthouse Records (Law Numbers 325748 and 34684) [Fairfax Courthouse]; *Wilmington News-Journal*, May 20, 1979.

Chapter 1: Interviews with Bagley Family Sources [Bagley]; Tennent Bagley Collection #1833, Howard Gotlieb Archival Research Center, Boston University [Collection 1833 BU]; David E. Hoffman, *The Billion Dollar Spy* (New York: Anchor Books, 2016) [*Billion*]; Maris Goldmanis, "The Case of Aleksandr Ogorodnik," Numbers Station Research Information Center; Martha Peterson, *The Widow Spy* (Wilmington, NC: Red Canary Press, 2012); Bob Fulton, *Reflections on a Life* (Bloomington, IN: AuthorHouse, 2008); Christopher Andrew and Vasili Mitrokhin, *The Sword and the Shield* (New York: Basic Books, 2001) [*Sword*]; Duane R. Clarridge, *A Spy for All Seasons* (New York: Scribner, 1977) [*Seasons*].

Chapter 2: Tennent H. Bagley, *Spy Wars* (New Haven, CT: Yale University Press, 2007) [*Wars*]; Bagley; Collection 1833 BU; *Widows*; Paisley; Raymond Rocca Collection #1832, Howard Gotlieb Archival Record Center, Boston University [Collection 1832 BU]; Tad Szulc, "The Missing CIA Man," *New York Times Magazine*, Jan. 7, 1961 [Szulc]; Internet Archives, Full Text of John Arthur Paisley, FBI, https:archive.org/stream/John Arthur Paisley [Internet Archives]; documents.theblackvault.com/documents/fbifiles/coldwar/johnpaisley.pdf [Black Vault]; US Senate Select Committee on Intelligence, Jan. 1, 1979, to Dec. 31, 1980, 97193.pdf [Senate]; Paisley Depositions; *Widows*; Paisley; Maryland State Police Report first cited in *Widows*; Joseph Trento, "The Spy Who Never Was," *Penthouse*, March 1979; Maryland Park Service Report, IR-45-78-268; Coast Guard Documents; CIA Security Memos, FOIA [Security]; Interviews with Intelligence Sources [IS]; *Wilmington News-Journal, New York Times, Washington Post* coverage [Press]; Colonial Funeral Home, Falls Church, Virginia, Records, first cited in *Widows* [Colonial Funeral]; Edward Jay Epstein, *The Annals of Unsolved Crime* (New York: Melville House, 2012) [*Annals*].

Chapter 3: Paisley; *Wars*; Tennent H. Bagley, *Spymaster* (New York: Skyhorse Publishing, 2015) [*Spymaster*]; Edward Jay Epstein, *James Jesus Angleton: Was He Right?* (New York: FastTrack Press, 2014) [*Right*]; David C. Martin, *Wilderness of Mirrors* (New York: Skyhorse Publishing, 2018) [*Wilderness*]; David Wise, *Molehunt* (New York: Random House, 1992) [*Molehunt*]; "Moles, Defectors, and Deceptions," Center for Security Studies Conference, Georgetown University, March 29, 2012, edited by Bruce Hoffman and Christian Ostermann [Conference]; Tennent H. Bagley, "Bane of Counterintelligence: Our Penchant for Self-Deception," *International Journal of Intelligence and Counterintelligence* 6, no. 1 (1993) ["Penchant"].

Chapter 4: Bagley; *Wars*; *Spymaster*; *Sword*; *Wilderness*; *Molehunt*; *Right*; IS; "Peter Deriabin, 71, a Moscow Defector Who Joined CIA," *New York Times*, Aug. 31, 1992; Peter Deriabin and Frank Gibney, *Secret World* (New York: Ballantine Books, 1980) [*World*].

Chapter 5: Bagley; IS; Collection 1832 BU; Jefferson Morley, *The Ghost* (New York: St. Martin's Press, 2017) [*Ghost*]; *Right*; Tom Mangold, *Cold Warrior: James Jesus Angleton: The CIA's Master Spy Hunter* (New York: Simon and Schuster, 1991) [*Cold Warrior*].

Chapter 6: Bagley; IS; *Right*; Conference; *Widows*, particularly interviews with Petty; *Molehunt*; *Wilderness*; *Wars*; Collection 1833 BU.

Chapter 7: *Widows*; Bagley; *Cold Warrior*; *World*; IS; *Ghost*; *Molehunt*; *Wilderness*; *Right*.

Chapter 8: *Wars*, particularly for quoted dialogue; *Wilderness*; *Ghost*; Bagley; IS; *Molehunt*; *Right*; Internet Archive Freedom of Information Act FBI Nosenko Files, identifier-ark: ark./13960/14dn92285 [FBI Nosenko]; *Sword*; William Hood, *Mole: The True Story of the First Russian Intelligence Officer Recruited by the CIA* (New York: W. W. Norton, 1982) [*First*]; Clarence Ashley, *CIA Spymaster* (Gretna, LA: Pelican, 2004) [Ashley]; John Limond Hart, *The CIA's Russians* (Annapolis, MD: Naval Institute Press, 2003) [*Russians*]; David E. Murphy, Sergei A. Kondrashev, and George Bailey, *Battleground Berlin: CIA vs. KGB in the Cold War* (New Haven, CT: Yale University Press, 1977) [*Battleground*]; *Seasons*.

Chapter 9: Bagley; *Wars*, particularly for quoted dialogue; FBI Nosenko; *First*; *Russians*; *Battleground*; *Wilderness*; *Ghost*; *Molehunt*; *Right*.

Chapter 10: *Wars*, particularly for quoted dialogue; Bagley; IS; *Russians*; *Ghost*; *Battleground*; *Cold Warrior*; *Wilderness*; Conference; Collection 1832 BU; Collection 1833 BU; "Penchant"; *Billion*.

Chapter 11: *Wars*; Bagley; House Select Committee on Assassinations, 95th Congress Hearings (Washington, DC: Government Printing Office, 1979) [Assassinations]; Report of the President's Commission on the Assassination of President John F. Kennedy (Washington, DC: Government Printing Office, 1964)[Report]; *Cold Warrior*; *Right*; Edward Jay Epstein, *Legend: The Secret World of Lee Harvey Oswald* (New York: McGraw-Hill, 1978) [*Legend*]; Press.

Chapter 12: *Wars*, particularly for dialogue; Assassinations; *Legend*; *Right*; *Ghost*; FBI Nosenko; Collection 1833 BU; Collection 1832 BU; Bagley.

Chapter 13: *Wars*, particularly his interview with Abidian; Bagley; *Ghost*; Report; *Sword*; John Barron, *KGB: The Secret Work of Soviet Secret Agents* (London: Hodder and Stoughton, 1974); Jerold L. Schechter and Peter S. Deriabin, *The Spy Who Saved the World* (New York: Charles Scribner's Sons, 1992) [*Saved*]; Oleg Penkovsky, *The Penkovsky Papers* (Garden City, NY: Doubleday, 1964); Ben Macintyre, *The Spy and the Traitor* (New York: Crown, 2019); "Penchant"; Conference; *Billion*; Leonard McCoy, "The Penkovsky Case," *Studies in Intelligence*, declassified September 2014; *First*.

Chapter 14: *Wars*, particularly for dialogue; *Ghost*; *Legend*; Report; Assassinations; IS; Bagley; Conference; "Penchant"; FBI Nosenko; *Sword*; *Cold Warrior*; "The Analysis of Yuri Nosenko's Polygraph Examination," Richard Arthur testimony to Select Committee on Assassinations, US House of Representatives, March 1979 [Polygraph]; *Wilderness*; *Molehunt*; Press.

Chapter 15: *Wars*; Bagley; Richard J. Heuer Jr., "Nosenko: Five Paths to Judgment," *Studies in Intelligence* 31, declassified Fall 1987 ["Paths"]; *Right*; *Ghost*; *Wilderness*; *Molehunt*; Polygraph.

Chapter 16: Bagley; IS; *Wars*; Collection 1832 BU; Collection 1833 BU; "Paths"; *Ghost*.

Chapter 17: Bagley; *Wars*; *Widows*.

Chapter 18: *Wars*; "Ex-CIA Employee Held as Czech Spy," *New York Times*, Nov. 28, 1984; Cynthia L. Haven, *The Man Who Brought Brodsky into English* (New York: Academic Studies Press, 2021); Tracy Burns, "Life During the Communist Era in Czechoslovakia," https:/www.private-prague-guide.com; Richard Cunningham, "How a Czech 'Super Spy' Infiltrated the CIA," *The Guardian*, June 30, 2016 [Cunningham]; Ronald Kessler, "Moscow's Mole in the CIA," *Washington Post*, April 17, 1988 [Kessler]; Ronald Kessler, *The Secrets of the FBI* (New York: Crown, 2011) [*Secrets*]; "Unknown Spy Sites," International Spy Museum, https://www.spymuseum.org ["Sites"]; Conference; *Sword*; Fairfax, Virginia, court records, cited first in *Widows* (at Law No. 38430) [Court]; *Billion*; *Right*; Press.

Chapter 19: "Sites"; Cunningham; Kessler; *Widows*, particularly Fairfax, Virginia, court records; Conference; *Wars*; Bagley.

Chapter 20: *Wars*; *Spymaster*; "Paths"; Cunningham; Kessler; Conference; *Billion*; *Sword*; *Widows*; Bagley; IS.

Chapter 21: *Wars*; Bagley; IS; *Spymaster*.

Chapter 22: Bagley; IS; Polygraph; *Widows*; *Molehunt*; *Ghost*; Bagley Testimony, House Assassinations Committee, Doc ID: 32273600; FBI Nosenko; "Paths"; *Legend*; *Right*; Szulc; William Safire, "Slithy Toves of CIA," *New York Times*, Jan. 22, 1979 [Safire].

Chapter 23: *Russians*; *Wars*; Bagley; John Steadman, "Forget ERA, Sivess' True Passion Was for CIA's Game of Intrigue," *Baltimore Sun*, March 17, 1966; *Widows*; *Spymaster*; *Ghost*; James Disette, "Cloak, Dagger, and Chesapeake," Parts I and II, https://www.chestertownspy.org/spies-pf-the-eastern-shore; Ann Hughey, "The House That Hid the CIA's Secrets," *Wall Street Journal*, April 19 1991; Polygraph; Collection 1832 BC.

Chapter 24: *Widows*, particularly Coast Guard records and family interviews; *Ghost*; *Cold Warrior*; Szulc; Safire; Internet Archives; Black Vault; Paisley; IS; Paisley CIA biographical file, FOIA[Biog File]; Bagley; IS; Press.

Chapter 25: *Wars*; *Widows*, particularly Paisley family interviews; Paisley; Deposition; Black Vault; Internet Archives; CIA, "Standard Assessment of Paisley," May 14, 1957, FOIA; Robert D. Vickers Jr., *The History of CIA's Office of Strategic Research* (Washington, DC: Center for the Study of Intelligence, 2019); David Shamus McCarthy, "The CIA & the Cult of Secrecy," 2008

Dissertation, William & Mary ScholarWorks [Dissertation]; Joe Trento, *Wilmington News Journal* and passim regarding P.O. box; Szulc; Safire.

Chapter 26: Thomas Powers, *The Man Who Kept Secrets: Richard Helms and the CIA* (New York: Knopf, 1979); *Wilderness*; Henry Hurt, *The Spy Who Never Came Back* (New York: Reader's Digest Press, 1983); *Widows*; Joseph J. Trento, *The Secret History of the CIA* (New York: Forum Prime, 2001); *Wars*; Bagley; Tad Szulc, "The Shadrin Affair: A Double Agent Double-Crossed," *New York Times Magazine*, May 8, 1978; *Sword*; Collection 1832 BC; "Paths"; Robert G. Kaiser, "A Non-Fiction Spy Story with No Ending," *Washington Post*, July 17, 1977; *Wars*; Bagley; IS; Press.

Chapter 27: *Wars*; Bagley; IS; Conference; "Penchant"; *Widows*; Kessler; Cunningham; Secrets; Court; Deposition; Internet Archives; Black Vault; Szulc.

Chapter 28: William Safire, "Deception Managers," *New York Times*, Aug. 6, 1981; Federation of American Scientists, "Weapons of Mass Destruction," https://nuke.fas.org/guide/russia/icbm/; The Special Collection Service & Gamma Guppy, https://nsarchive2.gwu.edu; *Widows*, particularly interviews with Richard Pipes and David Binder; Seymour Hersh, *The Price of Power* (New York: Summit, 1983); Dissertation; Jeffrey T. Richelson, "The CIA and Secret Intelligence," National Security Archive, March 2005; Jeffrey T. Richelson, *The Wizards of Langley* (New York: Basic Books, 2005); "The CIA Mission Impossible," *Time*, Feb. 6, 1978; Richard Pipes, "Team B: The Reality Behind the Myth," *Commentary*, Oct. 1986; Bill Keller, "The Boy Who Cried Wolfowitz," *New York Times*, June 14, 2003; Internet Archives; Black Vault.

Chapter 29: *Widows*; *Wars*; Bagley; *Russians*; IS.

Chapter 30: Deposition; *Widows*, particularly Betty Myers interview; Black Vault; Internet Archives; Bagley; "Paths"; House Assassination Committee Transcripts, 1978; Safire; Szulc; Fairfax Courthouse (originally cited in *Widows*); Paisley; Kessler; Conference; *Sword*; Jason Fangone, "The Amazing Story of the Russian Defector Who Changed His Mind," *Washingtonian*, Feb. 2, 2018; Stephen Engleberg, "CIA Gives a Rare Glimpse of Life of a Top Soviet Spy," *New York Times*, Nov. 9, 1991; *Annals*; *Right*; David Sullivan interview, cited in *Widows*.

Chapter 31: *Widows*, particularly David Sullivan interview and dialogue; Tim Weiner, "CIA Officer's Suit Tells Tale of Betrayal and Disgrace," *New York Times*, Sept. 1, 1996 [Weiner]; Richard K. Betts and Thomas Mahnken, editors, *Paradoxes of Strategic Intelligence* (Milton Park, UK: Routledge, 2004);IS; Dissertation; Internet Archives; Black Vault; Press.

Chapter 32: *Widows*, particularly David Sullivan interview and dialogue; Internet Archives; Black Vault; Weiner; Collection 1832 BC; Collection 1833 BC; Bagley; Paisley; *Annals*; Szulc; Safire; Coast Guard records, first cited in *Widows*; "Report to the Senate of Select Committee on Intelligence, January 1, 1979–December 31, 1980" (Washington, DC: Government Printing Office, 1981); Blaine Harden, "FBI Tells Senate Panel Paisley Probe Unjustified," *Washington Post*, Apr. 4, 1979; Timothy S. Robinson, "Full Report on Paisley to Be Secret," *Washington Post*, Apr. 24, 1980; Jesse Helms Archive Center, Record Group/Senatorial Papers, 1953–2004; *Wars*; Bagley.

Chapter 33: *Wars*, particularly for quoted dialogue with Kondrashev; Bagley; *Spymaster*; IS; *Saved*; "Penchant"; *Russians*; *Sword*; *Ghost*; *Molehunt*; *Legend*; Peter Wright, *Spycatcher* (New York: Penguin, 1987); Elaine Shannon, "Death of the Perfect Spy," *Time*, July 22, 2007; Hy Rothstein and Barton Whaley, *The Art and Science of Military Deception* (New York: Artech House, 2013).

Chapter 34: Bagley, particularly for passages that are specifically quoted and appear within quotation marks; IS; *Spymaster*; "Penchant"; Conference; "Paths"; David Robarge, "Cunning Passages, Contrived Corridors," *Studies in Intelligence* 53 (Dec. 2009); *Ghosts*.

Epilogue: Paisley; *Widows*, particularly for details about postcards and subsequent theft; Maryann Paisley, *Appellant v. Central Intelligence Agency*, 724 F.2d 201(D.C. Cir., 1984); Internet Archives; Black Vault; Szulc; Deposition.

Acknowledgments

A question often hurled my way, whether it's in the aftermath of a talk I've given on one of my books or at a dinner party among curious friends, is whether my job—or the work, for that matter, of any reporter who pokes around where he's not invited—is dangerous. In response, I usually muster up a stagy bravado and say my only real fear throughout all these years of writing and reporting has been of missing a deadline. But this book was different. I labored through a time of great personal stress. *The Spy Who Knew Too Much* was researched and written in a pandemic.

Now, one might think that the sedentary challenge of filling pages with words would not be too different in the time of COVID than it was in my previous world. And it wasn't. I'd be glued to my unforgiving desk chair, my thoughts drifting from the page to the views of ancient Connecticut fields and a murky pond, for hours on end. The outside world, with its unseen yet relentless and possibly hovering threat, seemed very distant from my hilltop.

However, researching this story was another matter. That's when I was required to venture out into an invisible, previously unimagined danger. Sure, following a path through the secret world is always a precarious business; the keepers of secrets are loath to share them. But this go-around the danger—a very real, almost palpable current of constant distress—was provoked neither by the intrusive questions I raised nor by my tenacious tapping the same spot in an attempt to get a substantive revelation. Rather, it was the circumstances of the interviews, the practicalities of the process, that gave me the willies.

One small example: It was the early days of COVID. Vaccines were still just a distant, and perhaps unlikely, promise. And I didn't dare fly; that seemed to be asking for trouble. Yet I had a source who, after a series of phone conversations, finally agreed to talk. He set a meeting at nine a.m. at an office complex along the Beltway outside Washington. So I hit the road in the still-dark wee hours of the appointed morning; I had no appetite for staying in a hotel. It was a long but uneventful drive, and I made it to the appointed destination with only minutes to spare.

And then it got menacing. I found myself seated across from my source in a windowless, and seemingly airless, conference room for hours. And, sporting a bewildering devil-may-care fatalism, he refused to wear a mask.

It was the most stressful interview of my career. I felt in constant danger.

And as my research led me deeper down the twisting path I navigate in this book, there were other situations filled with similar anxieties. Yet I more or less prevailed, escaped (up till now, at least) infection, and managed to huddle with sources willing to talk. And then I sat down to write this book.

So with gratitude and a measure of wonder that I was able to fulfill my mission in such perilous times, there are a few thank-yous that I want to offer to those on whom I counted on for advice and assistance.

Lynn Nesbit, my agent since the dawn of my career and my friend for all that time, too, is invaluable; she's wise, and candid, and always there when needed. Mina Hamedi in her office also has been generous with her kindness.

At HarperCollins, Jonathan Jao is not just a perceptive and thoughtful editor, but also a gentleman. And those qualities—rare gifts!—make it a delight to work with him on what has now become a growing stack of my books. Jonathan Burnham is another rarity in this business: a publisher who actually reads the books he brings into the marketplace, and I'm grateful for his support. And the rest of the HarperCollins team—David Howe, Tracy Locke, and Tom Hopke, among many others—have also been invaluable in helping to bring my books out into the world.

In Hollywood, Bob Bookman and I have worked together forever, or so it happily seems, and I look forward to our mutual toils (and frustrations) continuing for years to come. Jason Richman has become a smart and perceptive addition to this partnership. As is Craig Emanuel, whose wisdom, both legal and secular, I count on.

A large portion of the past couple of years has also been spent laboring to turn my previous book, *Night of the Assassins*, into a limited series. In this endeavor, I've had the good fortune to work with a very canny group who also, to my great enjoyment, happen to be gentlemen to boot: Alex Gansa, Howard Gordon, Glenn Geller, and, not least, the intimidatingly intelligent Jonathan Mostow.

And then there are the friends I counted on, even though the pandemic often kept us at a distance: Susan and David Rich; Irene and Phil Werber; John Leventhal; Bruce Taub; Betsey and Len Rappoport; Sarah and Bill Rauch; Barbara and Ted Ravinett; Pat, Bob, and Marc Lusthaus; Ken Lipper; Kathy Rayner; Claudie and Andrew Skonka; Nick Jarecki; Graydon Carter; Destin Coleman; Daisy Miller; Beth DeWoody; Arline Mann and Bob Katz; and Sara Colleton.

My three children—Tony, Anna, and Dani—are busily off in the world and their accomplishments fill their old father with great pride.

My sister Marcy is a godsend; I'd be lost without her.

And here on this hilltop, as we were locked down together while an invisible menace raged, Ivana made our life in isolation a joy.

Index

About the Author

HOWARD BLUM is the author of the *New York Times* bestseller and Edgar Award–winner *American Lightning*, as well as *Wanted!*; *The Gold of Exodus*; *Gangland*; *The Floor of Heaven*; *The Last Goodnight*; *In the Enemy's House*, which was a New York Times Notable Book; and most recently *Night of the Assassins*, which is being developed as a limited series by Sony Television. While at the *New York Times*, he was twice nominated for a Pulitzer Prize for investigative reporting. He is the father of three children and lives in Connecticut.